Geotextiles handbook

Geotextiles handbook

T. S. Ingold &
K. S. Miller

Thomas Telford, London

Published by Thomas Telford Limited, Thomas Telford House, 1 Heron Quay, London E14 9XF

First published 1988

British Library Cataloguing in Publication Data
   Ingold, T. S.
      Geotextiles handbook
      1. Synthetic fabrics in building
      I. Title   II. Miller, K. S.
      624.1′897   TA668

ISBN: 0 7277 1333 7

© T. S. Ingold and K. S. Miller, 1988

All rights, including translation, reserved. Except for fair copying, no part of this publication may be reproduced, stored in a retrieval system or transmitted in any form or by any means, electronic, mechanical, photocopying, recording or otherwise, without the prior written permission of the Publications Manager, Publications Division, Thomas Telford Limited, Thomas Telford House, 1 Heron Quay, London E14 9XF.

This guide is published on the understanding that the authors are solely responsible for the statements made and opinions expressed in it and that its publication does not necessarily imply that such statements and or opinions are or reflect the views or opinions of the publishers. Every effort has been made to ensure that the statements made and the opinions expressed in this publication provide a safe and accurate guide; however, no liability or responsibility of any kind can be accepted in this respect by the publishers or the authors.

Typeset in Great Britain by Katerprint Typesetting Services, Oxford.
Printed and bound in Great Britain by Redwood Burn Limited, Trowbridge, Wiltshire.

Contents

1. Introduction — 1
2. Materials and manufacturing processes — 4
   Woven fabrics; non-woven fabrics; knitted fabrics; meshes; grids
3. Properties and their measurement — 22
   Mass per unit area; pore size; surface friction; short-term tensile load-deformation characteristics; long-term tensile load-deformation characteristics; water permeability; durability; site quality control testing
4. General construction techniques — 37
   Site reception; storage; handling; laying; jointing; cutting; site damage and repair
5. Applications and construction techniques — 52
   Temporary roads; permanent roads; repair of permanent roads; railway tracks; embankments on soft ground; walls and steep sided embankments; drainage applications; revetments
6. Bibliography — 73
7. Compendium of product data — 75

1
Introduction

Geotextiles are thin, flexible, permeable sheets of synthetic material used to stabilise and improve the performance of soil associated with civil engineering works. Correctly designed and installed, geotextiles have the ability to filter, drain, reinforce and separate soil. In many applications, geotextiles may be designed and selected to perform a combination of these functions. For example, when installed at the base of a granular fill embankment constructed over soft clay all four functions might operate.

Like so many other techniques employed in civil engineering, the idea of using fabrics is not entirely novel. Over a century ago, sheets of canvas were incorporated in earth fill to reduce lateral thrusts exerted behind retaining walls. Some fifty years ago, cotton duck fabric, similar to denim, was used in the United States to stabilise dirt roads. In the early fifties, synthetic textiles were used by the Dutch as filters in the rapid repair of the North Sea dykes. Many more instances of this novelty can be found stretching back over the years; however, the development of modern geotextiles started in the late sixties.

In the post-war years there was massive restructuring in the European textile industry with much emphasis on the development of highly efficient production plant and the use of synthetics. In both Europe and the United States, domestic markets became more important, as many of the previously lucrative export markets were becoming self-sufficient. A prime example was the United Kingdom cotton industry, which in the pre-war years supplied massive markets throughout the world. These markets dwindled rapidly after the war and in some cases former importers became exporters of textile products. The western textile industry came to terms with these problems, and during the boom years of the sixties the expansion of the textiles market was matched by an expansion in capital investment and technology. Towards the end of the decade, domestic markets were progressively eroded by cheap imports and a decline in demand, thereby leaving western producers with excess production capacity and heavily committed investments.

As a consequence, no doubt, of extensive market research, the textile manufacturers lighted on the civil engineering industry and its enormous potential for consumption. Of particular interest were the so-called 'high volume markets' of roads and railways, which even today account for a very large proportion of geotextile sales. During their initial development of the market, textile manufacturers poured vast resources into promotion and research. Although the textile industry was castigated by some sceptics for its commercialism, it is unlikely that geotextiles would have developed without such impetus and resources. It is fair to say, however, that in the early days of development the civil engineer needed a healthy scepticism, especially when it became apparent that many geotextiles were identical to fabrics sold as carpet backing or upholstery lining. Since these early days, geotextiles technology has made huge advances. However, there is still more to be done to develop a sound design technology covering all fields of application.

A turning point in geotextiles technology came in 1977 with the *International Conference on the Use of Fabrics in Geotechnics* held in Paris. With hindsight, this can be regarded as the first international conference on geotextiles. The second and third international conferences on geotextiles were held in Las Vegas in 1982 and Vienna in 1986, respectively.

In line with many other international conferences, geotextiles conferences have settled into a four-year cycle, with the next conference scheduled to be held in The Hague in 1990. Four years is a long time in a rapidly expanding technology and, consequently, there are an increasing number of national and regional conferences appearing, as well as two regular technical publications. The first of these is *Geotechnical Fabrics Report* published bimonthly by the Industrial Fabrics Association International in St Paul, Minnesota. In contrast, the *International Journal of Geotextiles and Geomembranes* is a learned journal published quarterly by Elsevier Applied Science Publishers of London.

Since geotextiles are truly international products, it is import-

Introduction

ant that unified testing standards and definitions are developed. There are at present several national standards issued or under development, including a British Standard for index testing of geotextiles. Work is already in hand on an international standard for testing under the auspices of the International Organisation for Standardisation (ISO). However, it is likely to be some considerable time before an international standard comes to fruition. In parallel with the various national and international committees considering testing, the profession is organised on a wider basis through many national geotextile societies, and on an international basis through the International Geotextile Society (IGS), which was formed in 1983. The British chapter of this society was formed in 1987.

2
Materials and manufacturing processes

The properties of a textile will be radically affected by the material of the textile and the structure of the textile imparted by the manufacturing process. For example, sack cloth is commonly made of woven jute or hemp. A potato sack produced from this cloth is strong and of low extensibility. If other materials and manufacturing processes were employed, an unsatisfactory product may result. For example, a knitted lambswool potato sack would not prove very serviceable. Even a woven hemp sack can become unserviceable if stored in a damp or wet area, as it will rot. This tendency to rot in damp environments is one of the reasons that natural fibres such as jute or cotton are not normally employed in the manufacture of geotextiles, except in applications such as erosion control where biodegradability is an asset. Instead, use is made of plastics.

Plastics are synthetic organic materials, based on carbon, and are generally obtained by a chemical process. A common process is the distillation of crude oil. There are two very important divisions in plastics: the thermoplastics and the thermosetting plastics. The thermoplastics can be softened and rehardened by heating and cooling in much the same way as candle wax. Thermosetting plastics are quite different, because once they harden from their initial molten state, they cannot be resoftened by the application of heat—like a hard boiled egg which no amount of reheating will convert to its original liquid phase.

Thermoplastics are the raw material of geotextiles. The fundamental properties of a particular thermoplastic will depend on its structure, any additives and the process by which it is formed. The basic unit of any thermoplastic is the monomer which is a molecule of several atoms. One of the most simple is the hydrocarbon ethylene monomer shown in Fig. 1. By various chemical

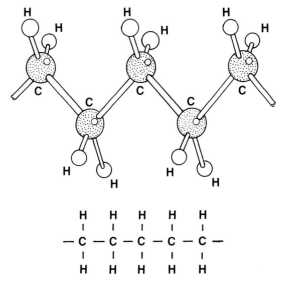

Fig. 1. Ethylene monomer

Fig. 2. Polyethylene molecule

processes the monomers can be joined to form long molecular chains. The process is called polymerization and the resulting product is a polymer. Fig. 2 shows several ethylene monomers joined to form a polyethylene molecular chain. With the inclusion of various fillers and chemical additives, these chains form a particular thermoplastic. Many different polymers may be used to manufacture geotextiles: for example, polyamide (Nylon), polyester (Terylene), polyvinyl chloride (PVC), polypropylene or polyethylene. The last two polymers belong to a very similar hydrocarbon family, the polyolefins. In addition, many forms of polyester are used, of which polyethylene-terephthalate is the most common. By far the most widely used materials in the manufacture of geotextiles are the polyesters and the polyolefins. Each raw polymer has its own strengths and weaknesses; for example, polyester has great strength and resistance to creep, while the polyolefins have good resistance to attack by organic acids. These are characteristics of the polymer which can be altered to a greater or lesser degree by various chemical additives. Improvements can also be made by mechanical means, such as drawing, during the manufacturing process of the components which are ultimately converted into the geotextile. Drawing or stretching of the polymer will tend to align the polymer chains which otherwise take up a minimum energy configuration, rather like a coil spring. This process of orientation, generally undertaken when the polymer is warm, will straighten the coiled polymer chain, thus reducing strain and increasing strength in the finished product. With all these processes, care must always be taken to ensure that enhancement of one desired property is not at the detriment of some other property.

In general, the manufacturing process of a geotextile comprises at least three stages: the production of the polymer with its various additives; the production of a component; and the conversion of the component into the finished geotextile. The polymer is generally made in a chemical processing plant and is supplied to the manufacturer in the form of pellets or granules which are reheated for conversion—usually in two stages—into the geotextile. The first of these two stages is the formation of a basic component. The physical forms of these components can vary but they generally fall into one of three broad categories

- (*a*) a continuously extruded circular cross-section filament having a diameter generally measuring a fraction of a millimetre, and an indefinite length
- (*b*) a continuously extruded flat tape having a breadth of several millimetres, a fraction of a millimetre thickness, and an indefinite length
- (*c*) an extruded sheet or film of width up to several metres and thickness varying from a fraction of a millimetre (film) to several millimetres (sheet).

At this stage the civil engineer is faced with perhaps half a dozen polymers which have been converted into one of three basic components. These components are then converted into the finished product with perhaps some intermediate processing of the basic component before final conversion into the finished product. The manufacturing processes used for this conversion can result in many different geotextile structures. Since each particular structure, and the polymer used to form this structure, will control the mechanical properties, hydraulic properties and durability of the geotextile, it is useful to list these structures before giving more detailed consideration to the manufacturing processes leading to their formation. The basic structures are: woven fabrics, non-woven fabrics, knitted fabrics, meshes and grids.

The three basic components used to make these structures are often subject to subsidiary processing before final conversion. For instance, continuous filaments can be twisted into yarn, aligned into parallel groups to form a multifilament, cut into short lengths—typically 50 mm long—to form 'staple fibre', or the filament can be used alone as a monofilament. Similarly, the flat tapes, formed either by direct extrusion or by slitting extruded film, can be used directly or twisted together to form a tape yarn. In the latter case, a single wide tape may be used, and this is often nicked, with short discontinuous cuts running down the length of the tape, to produce a fibrillated tape. In the manufacture of extruded tapes, it is quite common practice to stretch or draw the tapes directly after extrusion to align or orient the polymer molecules, thereby giving higher strength and axial tensile stiffness.

Woven fabrics

As the name implies, woven fabrics are obtained by conventional weaving processes, using a mechanical loom. In this process, an array of parallel elements—the warp—is beamed into the loom, and transverse elements—the weft—are threaded over and then under alternate warp elements. The woven product emerges from the loom and is wound into rolls. The type of weaving described is plain weave, of which there are many variations, such as twill, satin and serge; however, plain weave is the one most commonly used in geotextiles. The elements used are either flat tapes (extruded or slit film), tape yarn, multifilaments (generally as low twist yarn), or single monofilaments. Resulting structures are typically one millimetre thick with a comparatively regular distribution of pore or mesh openings which vary in dimension over a reasonably small size band. Wovens may be made up entirely of one element—for example, a tape warp and tape weft—or of two element types, one warp and one weft. The description of the structure would then be abbreviated—for example, a 'tape-on-tape' fabric, or if the warp is a monofilament and the weft a tape the structure would be a 'monofilament-on-tape'. Typical woven structures are illustrated by the photomicrographs in Figs 3–9, which give an indication of element sizes and pore sizes. There are no rigid criteria relating polymer type to structure; however, tapes are most commonly polypropylene, and monofilaments are most commonly polyethylene, whereas the finer multifilaments or multifilament yarns are commonly polyester.

Non-woven fabrics

Again, as the name implies, these structures are obtained by processes other than weaving. In the case of non-wovens, continuous monofilaments are usually employed; these may, however, be cut into short staple fibres before processing.

The first step in processing involves continuous laying of the fibres or filaments on to a moving conveyor belt to form a loose

Fig. 3. *Monofilament-on-monofilament geotextile*

Fig. 4. Monofilament-on-multifilament geotextile

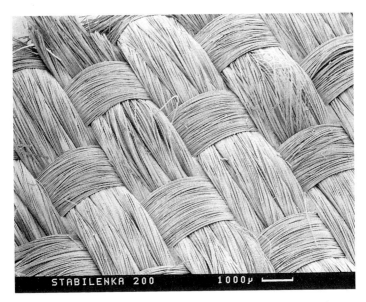

Fig. 5. Multifilament yarn-on-multifilament geotextile

Fig. 6. Monofilament-on-tape geotextile

Fig. 7. Slit tape-on-fibrillated slit tape geotextile

Materials and manufacturing processes

Fig. 8. Extruded tape-on-extruded tape geotextile

Fig. 9. Fibrillated tape yarn-on-fibrillated tape geotextile

web slightly wider than the finished product. This passes along the conveyor to be bonded. The bonding process used falls into one of three broad categories: mechanical bonding, thermal bonding or chemical bonding.

Mechanical bonding

Mechanical bonding is achieved by passing the loose web beneath a bank of reciprocating barbed needles which penetrate the full thickness of the web. As the needle enters the web, it drags some of the filaments, or staple fibres, down into the body of the web, causing them to interlace with other filaments. Each needle table contains many thousands of needles, and by adjusting the intensity of distribution of these needles, it is possible to control the density and compactness of the finished product. The resulting fabric is aptly termed 'needlepunched'. An example of the structure obtained by needling is indicated by the photomicrograph in Fig. 10.

Thermal bonding

Thermal bonding imparts cohesion to the web by fusion of the

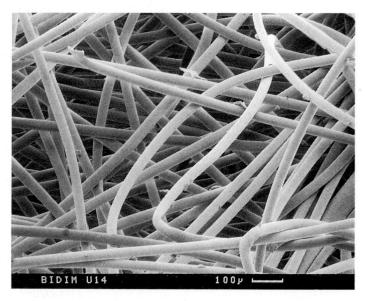

Fig. 10. Needlepunched continuous filament geotextile

continuous filaments at their cross-over points. Although these filaments are all of similar physical form, filaments of slightly different polymer chemistry may be employed to give different melting points.

The preoccupation with melting points of the polymers becomes apparent when consideration is given to the techniques used in achieving the thermal bond. These involve either passing the web through hot compressive rollers or passing it through a linear oven. In either case, the objective is to achieve sufficient melt to cause fusion of the filaments at their points of contact. Clearly, this can be a precarious process if all the filaments have the same melting point. Perhaps one of the most ingenious solutions to these potential problems is the heterofilament. This is a bi-component filament with a core of high melting point polymer surrounded by an outer coaxial sheath of lower melting point polymer. A heterofilament web can be passed through the linear oven, and only the outer sheath will melt to form the bond. The structure obtained using these processes is indicated by the photomicrograph in Fig. 11.

It is difficult to categorise many of the structures of the various products available, and this applies to the structure shown in Fig. 12. This is often referred to as a 'mat' and comprises an unconsolidated thermally bonded web of very large diameter monofilaments. The filaments used are of the order of one millimetre in diameter compared with the few tens of microns for the filaments in the fabric shown in Fig. 11. Mats can be many millimetres thick and can have a very open structure. This is revealed by comparing the scales of Figs 11 and 12.

Chemical bonding

Chemical bonding generally follows a needling process and involves imparting further cohesion to the web of filaments, or staple fibres, by the addition of a chemical binder. Most commonly, the binder is acrylic and may be applied by total immersion, or dipping, of the fabric in a bath, or by spraying. Following the application of the binder, curing is achieved by passing the treated fabric through a linear oven or hot rollers. In some processes the binder is actually applied by a printing technique. A typical structure is illustrated in Fig. 13.

Geotextiles handbook

Fig. 11. Heat bonded continuous filament geotextile

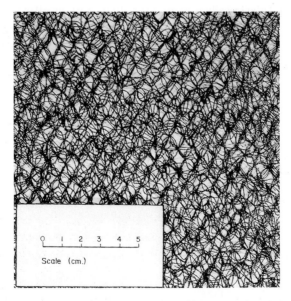

Fig. 12. Thermally bonded mat structure

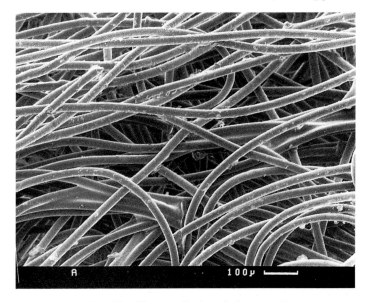

Fig. 13. Chemically bonded geotextile

Knitted fabrics

At present, simple knitted structures have no apparent application as geotextiles, as they tend to suffer excessive elongation under tension. However, the technique of warp knitting can be employed to impart high unidirectional strength. This process involves laying high strength multifilament yarns into an extensible knitted base, which acts as a carrier or substrate for the high strength component. The end result is a fabric which exhibits high strength and low elongation in the longitudinal (warp) direction. The structure of the one known geotextile product in this category is illustrated by the photomicrograph in Fig. 14.

Meshes

Meshes are not textile fabrics; however, as they can perform functions associated with geotextiles, they may be deemed geotextiles. In essence, meshes have openings or pores which are larger in dimension than the two sets of members which combine to form the mesh. For civil engineering applications, meshes are produced by a process of continuous integral extrusion, using a simple but

Geotextiles handbook

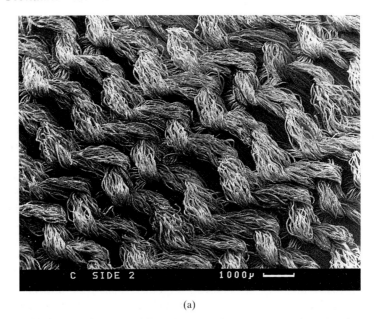

(a)

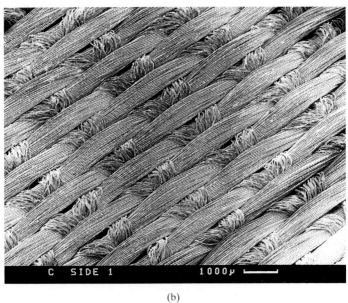

(b)

Fig. 14. Knitted geotextile: (a) knitted base; (b) upper surface

ingenious rotating die. A much simplified section of the die is shown in Fig. 15, from which it can be seen that the die consists of an inner disc with longitudinal slots around its outer periphery and an outer annulus having longitudinal slots around its inner periphery. When the die is stationary, it follows that extrusion of plastic through the die would produce strands or members having the same cross-section as the slots. If, however, the outer die is rotated clockwise against a plain inner die, the extrusion would be a series of unconnected strands forming a clockwise spiral. Similarly, counter-clockwise rotation of a slotted inner die against a plain outer die would form a counter-clockwise spiral extrusion. When both inner and outer slotted dies are counter rotated (Fig. 15), then two series of spiral extrusions are formed simultaneously; however, when the slots in the two sections of the die come into alignment, there is, for a short period of time, only one set of strands extruded, thereby joining the two spiral extrusions together with a series of integral junctions to form a tubular mesh. The mesh opening is generally of diamond shape, and if the tubular sleeve is slit open on its long axis directly after extrusion, it can be flattened to form a continuously extruded sheet of diamond shape mesh. Before splitting, the tubular mesh is generally passed over a mandrel to expand its diameter and change the shape of the mesh opening before the slitting process is employed to produce flat sheet. Another variation on the process is to slit the tubular extrusion on the bias, parallel to the direction of one set of spiral members, to form an orthogonal grid. An example of the structures obtained in these unorientated meshes is illustrated in Fig. 16.

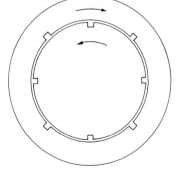

Fig. 15. Rotating die

Geotextiles handbook

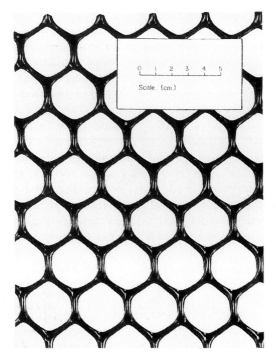

Fig. 16. Extruded mesh

Grids

Grid structures, often termed geogrids, are characterised by openings which can be larger in dimension than the sets of members making up the solid component of the grid. Pseudo textile grid structures can be formed using special weaving techniques such as leno weave, which produces large orthogonal pores, or by heat bonding two orthogonal sets of strands or tapes. The method employed for the production of Tensar grids involves a patented method of processing sheet polymer. Two or three stages are involved in the manufacturing process, which is illustrated diagrammatically in Fig. 17. The first stage involves feeding a sheet of polymer, several millimetres thick, into a punching machine, which punches out holes on a regular grid pattern. Following this, the punched sheet is heated and stretched, or drawn, in the machine direction. This distends the holes to form an elongated grid opening. In addition to changing the initial geometry of the holes, the drawing process orients the polymer molecules in the direction of drawing. The degree of orientation will vary along the length of the grid; however, the overall effect is an enhancement of tensile strength and stiffness. The process may be halted at this stage, in which case the end product is a uniaxially orientated grid. The resulting structure is illustrated in Fig. 18. Alternatively, the uniaxially orientated grid may proceed to a third stage of processing to be warm drawn in the transverse direction, in which case a biaxially orientated grid is obtained. In this structure the grid opening is very nearly square, as illustrated in Fig. 19. Although the temperatures used in the drawing process are above ambient, this is effectively a cold drawing process, as the temperatures are significantly below the melting point of the polymer.

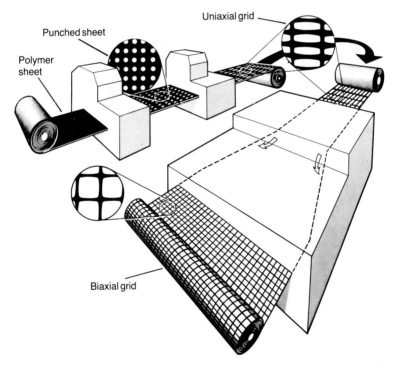

Fig. 17. 'Tensar' manufacturing process (Courtesy Netlon Limited)

Geotextiles handbook

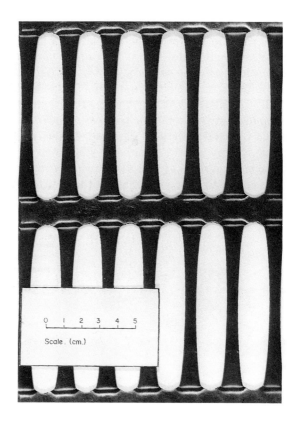

Fig. 18. Uniaxial grid

Materials and manufacturing processes

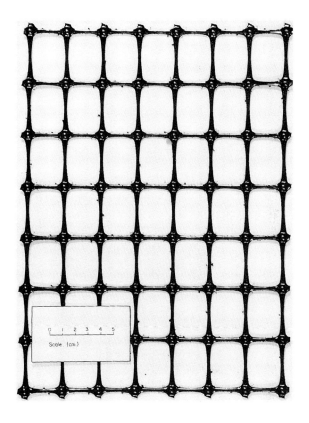

Fig. 19. Biaxial grid

3
Properties and their measurement

In civil engineering applications, the main geotextile properties of interest are the physical, mechanical and hydraulic properties. These will largely be a function of the polymers used and of the manufacturing process which creates the structure of the textile. Great care must be taken in quantifying these properties by testing, as the perceived properties of geotextiles, somewhat like soils, are a function of the test method and procedure employed. It is for this reason that much effort is being expended in developing standards that will lead to consistency in testing. In assessing geotextile properties, due regard must be given to the effects of degradation. Geotextiles, in common with all building materials, are prone to a progressive deterioration in properties. The rate and severity of this degradation are dependent on many factors, such as the environment, the polymer of the geotextile and the state of stress in the geotextile. Perhaps one of the most publicised modes of degradation is irradiation by ultra-violet light. However, as geotextiles are usually not exposed to the sun once installed, any potential problems can usually be avoided by planning construction to minimise exposure.

Test methods for geotextiles are far from being finalised, as international and many national standards for testing are still undecided. At present, there is no universally accepted code for the testing of geotextiles. Consequently, several different test methods may at present be in use to determine a given geotextile property. In view of this, any test methods mentioned here are illustrative rather than definitive, and only a limited number of fundamental geotextile properties are considered (see Table 1). It is these basic properties which are often used as the yardstick for site based quality control.

Some indication of the wide range of geotextile properties is

Properties and their measurement

Table 1. Basic geotextile properties

Physical	Mechanical	Hydraulic	Durability
			Resistance to degradation by:
Mass/Unit area	Short-term load/deformation	Permittivity	Irradiation Chemicals
Pore size			Bacteria Abrasion
Surface friction or adhesion	Long-term load/deformation (creep)	Transmissivity	Tearing Puncturing

given in the Appendix, which lists products that are readily available in the United Kingdom.

Mass per unit area

A basic physical property of geotextiles is mass per unit area expressed in units of grams per square metre: g/m^2. For commonly used geotextiles, mass per unit area varies in order of magnitude from typically 100 g/m^2 to 1000 g/m^2. The lighter weight fabrics, typically less than 250 g/m^2, tend to be the woven tape, thin needlepunched or heat bonded fabrics. The heavier weight fabrics generally comprise the thick needlepunched felts, woven tape yarn or some of the heavyweight woven polyester soil-reinforcing fabrics. Although not fabrics, most meshes and grids are generally found to fall into this heavyweight category. Although an apparently rather banal physical property, mass per unit area can be a good indicator of other properties and price. For example, in the case of the polyolefins, a controlling factor in price is the weight of the raw polymer contained in the fabric, because the manufacturing process (provided that it does not involve expensive additives or resins) is a comparatively fixed additional cost. Therefore, a plot of geotextile price against mass per unit area often defines a very narrow band which can show up products where a particularly high profit is included in the asking price. There is often a close correlation between unit weight and the strength of geotextiles of a given structure and polymer. This is illustrated in Fig. 20, which shows relationships between tensile strength and mass per unit area. It should be noted that strength is quoted in units of kilonewtons per metre width: kN/m. Other strength properties will

Geotextiles handbook

also vary with mass per unit area—for example, tear strength, which is the tensile force required to propagate tearing in an initially torn fabric.

Similarly, the force required to push a California Bearing Ratio (CBR) plunger through a constrained geotextile sample will increase with mass per unit area, thereby quantifying the intuitive assessment that heavier geotextiles are more resistant to puncturing and bursting. Increasing mass per unit area can, for certain non-woven structures, be associated with a decrease in properties

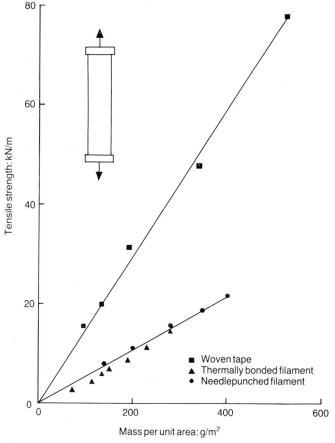

Fig. 20. Variation of tensile strength with mass per unit area for polypropylene geotextiles tested to DIN 53857

such as characteristic pore size and the ability to transmit water in a direction normal to the plane of the geotextile. Although not directly related to mass per unit area, an important factor can be the relative density of the fabric and polymer. For example, polypropylene floats while polyester does not; therefore, the use of polypropylene fabrics for certain submarine works can prove difficult.

Pore size

Two vital factors in the performance of geotextiles as filters are the size and distribution of the holes or pores in the geotextile. As can be seen from the photomicrographs in Figs 3–11 and 13–14, all geotextiles have pores which allow the transmission of water through the geotextile. When the geotextile is used as a filter, there is a need to allow the transmission of water while at the same time preventing the continuous transmission of soil particles. This can be achieved by selecting geotextiles which have pore sizes related, by a design method, to the particle sizes and coefficient of uniformity of the soil to be retained. The pores in a given geotextile are not of one size but are of a range of sizes, and the pore size distribution can be represented in much the same way as the particle size distribution for a soil.

Examples of pore size distributions for a heat bonded non-woven and a plain woven geotextile are given in Fig. 21. As can be seen from Fig. 21, and the photomicrographs in Figs 3–9, the pores in a woven fabric tend to be fairly uniform in size and regularly distributed. In contrast, Fig. 21 and the photomicrographs in Figs 10–11 show the pore sizes of non-wovens to vary over a wider range and to be more randomly distributed over the surface of the geotextile. To aid filter design, geotextile performance is commonly characterised by the O_{90} pore size (Fig. 21). This is determined directly from the pore size distribution curve, which itself is determined by sieving various known sizes of sand or glass beads (ballotini) through the geotextile. At present, there is no internationally agreed method for determining the characteristic pore size, as a variety of different wet or dry sieving techniques are employed. Indeed, standardisation of the particular pore size taken to characterise filtration performance has not yet been achieved. In consequence, either the O_{90} or the O_{95} pore size, or the effective opening size (EOS) may be quoted.

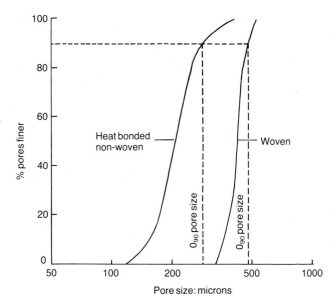

Fig. 21. Pore size distributions of typical geotextiles

For a given geotextile the perceived pore size will vary with test method, and, therefore, different manufacturers' data may not be directly comparable. It is important to bear this in mind in filter design, as any particular design may relate to only one type of test. All pore sizes should be quoted in microns (μm) rather than in sieve sizes. The range of O_{90} pore sizes varies from a maximum of approximately 1000 μm for a coarse woven monofilament-on-monofilament to a few tens of microns for the thick non-woven fabrics. In general, non-wovens exhibit smaller O_{90} pore sizes than wovens; however, there is a degree of overlap in the commonly employed O_{90} sizes, which vary from approximately 50 μm to 350 μm for the non-wovens and from 150 μm to 600 μm for the wovens.

Surface friction

When a geotextile is used as soil reinforcement, it is important that the bond developed between the soil and the geotextile is sufficient to stop the soil from sliding over the geotextile or the geotextile from pulling out of the soil when the tensile reinforcing

Properties and their measurement

load is mobilised in the geotextile. Sliding resistance between soil and geotextile is primarily a function of the surface roughness of the fabric. There can be a certain enhancement of bond when the pore sizes of the fabric are compatible with the particle sizes of the soil, in which case the soil can embed itself in the surface of the fabric. Sliding resistance is commonly assessed using a shear box, with the geotextile mounted on a rigid block in the lower half of the box (Fig. 22). Soil is placed in the upper half of the shear box, and the variation of shear stress against normal stress is determined in the same manner as for a conventional shear box test. For granular soils, tests are first carried out using the shear box filled with soil to determine the internal angle of shearing resistance (ϕ') of the soil alone. The procedure is repeated with the geotextile in the lower half of the shear box to determine the angle of bond stress (δ) betwen the soil and geotextile. For most geotextiles, the ratio δ/ϕ' rarely drops below 0·75 and is often close to unity; however, there are exceptions. For example, Fig. 23 shows the results obtained for a warp knitted fabric which has one

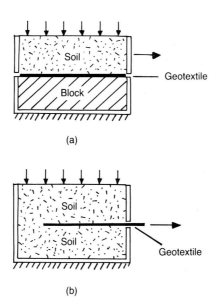

Fig. 22. Surface friction tests: (a) fixed shear box test; (b) pull-out test

Geotextiles handbook

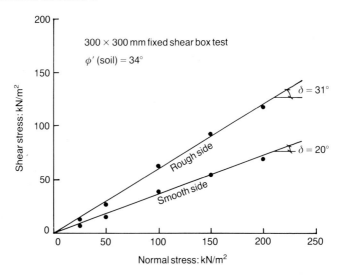

Fig. 23. Test results for knitted/woven geotextile

rough side and one much smoother side. The measured values of δ/φ' were 0·91 and 0·59 respectively. Although the average value is 0·75, it is clear that the accidental use of this fabric with the smoother face against the overlying soil mass to be retained could severely reduce the calculated factor of safety based on an assumed value of δ/φ' of 0·91. Should the problem under consideration involve undrained sliding of clay over a geotextile, then similar test techniques may be employed. However, the adhesion generated between the clay and geotextile is generally quoted as a proportion (α) of the undrained shear strength of the clay, where α is an adhesion factor similar to that employed in the design of piles in clay soils.

Pull-out resistance can be calculated from the results of shear box tests, provided that the extensibility of the geotextile is taken into account. A qualitative assessment of pull-out resistance can be made using a pull-out test (Fig. 22). Although great care is needed in attempting a quantitative interpretation of test results for extensible geotextiles, the pull-out test has the ability to detect faults in the structure of the geotextile, which would most likely go undetected in the shear box test. Some indication of the range of results that can be obtained is given in Fig. 24.

Properties and their measurement

Short-term tensile load-deformation characteristics

Constant rate of strain axial tensile tests give an indication of the tensile strength of a geotextile (measured in kN/m) and of the axial strain at rupture. However, the measured strength and particularly the rupture strain are a function of many test variables, including sample geometry, strain rate, temperature and the amount of any normal confinement applied to the geotextile. If a narrow sample is used (as in the 50 mm wide strip test, where the sample has a 200 mm gauge length) there will be excessive necking of the sample which exaggerates axial strains. This test is totally inappropriate in relation to geotextiles, because, when embedded in the soil, geotextiles subject to tension suffer elongation under plane strain conditions in which no necking occurs.

The importance of strain rate and temperature stems from the fact that all polymers are visco-elastic materials; therefore, when a load is applied, an elastic strain and a creep strain will result. When a low strain rate is applied, the sample takes longer to come to failure and, therefore, the creep strain is greater. In consequence, high rates of strain (which can be as much as 100% per

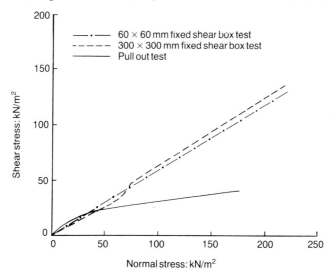

Fig. 24. Comparison of test results for heat bonded non-woven geotextile

minute) tend to produce lower failure strains and sometimes to yield higher strengths than do low rates of strain. Temperature has the reverse effect, with higher temperatures being associated with higher creep rates and, therefore, larger strains to rupture. To minimise these problems, it is recommended that a wide test sample should be used with a width-to-gauge length ratio of at least two, and that testing should be carried out at a standard temperature.

Regarding the effects of normal pressure, it is found that in many non-woven fabrics the rupture load is increased and the failure strain decreased if the geotextile is subject to a confining stress normal to its plane. Clearly, this would occur in practice, as the geotextile would be confined by an overburden of soil. The mechanism of this enhancement is simply the frictional force developed between the filaments of the fabric. Despite this enhancement, non-woven fabrics are, weight-for-weight, weaker and more extensible than woven fabrics, which are little affected by confinement. Woven fabrics tend to be strong and stiff, with tensile strengths up to 1000 kN/m and rupture strains as low as 12%. In contrast, non-woven fabrics tend to have tensile strengths in the range 5–20 kN/m, with rupture strains often around 50%. Rupture can occur more readily in certain fabrics when they are wet; in which case, if the application in mind involves stressing fabrics when wet, samples should be pretreated by soaking (usually for twenty-four hours) before testing.

Long-term tensile load-deformation characteristics

The results of short-term constant rate-of-strain tensile tests give a reasonable assessment of rupture strength and strain which might be used to select a suitable geotextile to withstand the rigours imposed by certain placing techniques such as rope-hauling. The same results might be applied to low-risk reinforcing applications such as unpaved roads. However, because such tests give little indication of how rupture loads or strains change with time, they have no use in the design of reinforced structures such as steep-sided embankments or vertical faced reinforced soil walls. In these long-term applications, the reinforcement must be designed to withstand a tensile load both without rupture and without the strain in the geotextile exceeding a predetermined limit. Of these two, the prime criterion is that there should be no tensile rupture of the reinforcement, as this would result in collapse of the structure. The essence of long-term behaviour can be explained with the aid of Fig. 25, which shows a family of conventional creep curves with each curve pertaining to a different constant applied tensile load. The sample at the highest loading will come to ductile failure first, with the time to ductile creep failure increasing as the applied load decreases. In Fig. 25, the different constant load applied to each of the five samples is shown in absolute terms. This load is usually also defined as a percentage of the short-term rupture load, although this will be meaningless unless the specification for the short-term test is quoted. One of the most important features of the conventional creep plot of total strain against time (with time usually to a logarithmic scale) is that, at any given time, total strain is the greatest for the most heavily loaded sample. Also, it may be found that, for any given time, the *strain rate* is the highest for the most heavily loaded sample. It is always the case that, whatever the constant load intensity, strain increases with time.

It is imperative that the structure does not collapse, and therefore a prime criterion is that the required restoring forces should never at any time exceed the factored creep rupture strength of the geotextile. From the point of view of serviceability, it is important that excessive movements do not occur in the

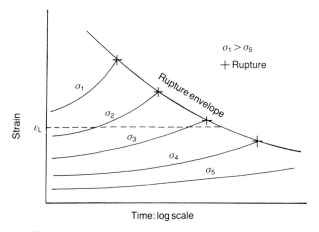

Fig. 25. Generalised strain against log time plot

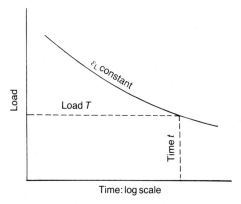

Fig. 26. *Stress-relaxation plot*

geotextile during the design life of the structure. To prevent this, the design of the structure must take long-term strain into account, which might be achieved by limiting strain to some predetermined value, say ε_L (Fig. 25). In theory, total strain can be maintained at a constant value by *decreasing* the applied tensile load with time. To see how load in a geotextile decreases with time, it is possible to strain samples to a predetermined strain and measure the variation in load with time. A time-dependent reduction in load under constant strain is called 'stress relaxation'. If loads are plotted against their respective times, a stress-relaxation plot results for the given constant limit strain ε_L (Fig. 26). To limit ductile creep strain to the value ε_L, after a time t, the applied load may be limited to a value T (Fig. 26).

Water permeability

A geotextile is, by definition, permeable, and it therefore exhibits the ability to transmit water in a direction normal to its plane. In general, this facility cannot be quantified by a unique coefficient of permeability for a given geotextile; to do so would presuppose laminar flow in which the mean flow velocity is directly proportional to hydraulic gradient, the constant of proportionality being defined as the coefficient of permeability of the conducting medium. It cannot be assumed that flow through geotextiles obeys Darcy's law or that permeability is a constant. This stems from the fact that flow normal to the plane of the geotextile may not be laminar and, in the case of non-wovens, that thickness, porosity and, therefore, any coefficient of permeability decrease with increasing normal pressure.

In consequence, it will be found that the ability of a geotextile to transmit water in a direction normal to its plane is expressed in a number of ways. Perhaps the most straightforward is the so-called 'water permeability' which is the flow rate per square metre through a geotextile for a defined head of water. The units are litres per second per square metre (l/s per m^2) for a driving head of either 5 cm or, more usually, 10 cm of water. As with permeability, this will be affected by the stress level as indicated in Fig. 27. Most non-woven geotextiles exhibit high water permeability under zero stress, say typically 200 l/s per m^2 for a 10 cm head. However, this drops approximately an order of magnitude for any significant stress level. Woven geotextiles are much less affected by stress level, but their water permeability is dramatically controlled by the structure of the fabric. The common, and generally less expensive, tape-on-tape fabrics have a low open area ratio and, in consequence, exhibit water permeabilities typically in the range 10–30 l/s per m^2 for a 10 cm head. In contrast, the woven monofilament-on-monofilament fabrics have much larger open area ratios, giving water permeabilities in the range 100–1000 l/s per m^2 for a 10 cm head.

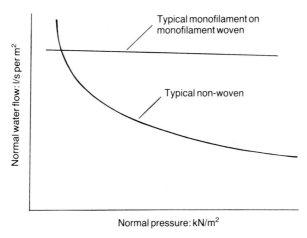

Fig. 27. Variation of flow rate with normal pressure

Durability

The durability of a geotextile may be regarded as the ability to maintain with time a required degree of integrity. This, of course, presupposes that the geotextile has been selected on the basis of its being sufficiently robust to endure the rigours of handling and installation. Once installed the selected geotextile should have adequate resistance to attack. This involves engineering and selecting the geotextile to resist the modes of attack associated with the particular service environment. If there is some inevitable form of degradation then the rate of degradation might be assessed with a view to allowing for it in the design.

Geotextiles may be subject to many forms of degradation such as mechanical or chemical attack. Examples of mechanical attack are abrasion, as might occur beneath badly installed armour stone subject to wave action, or attack by burrowing animals or insects. Chemical attack can be initiated in many forms; direct chemical attack may occur in very acid or alkaline soils. Alternatively, a geotextile might be used in a storage lagoon receiving highly active wastes which, if allowed to contact the geotextile, would lead to chemical attack. Although no direct chemical action is involved, certain chemical environments can induce environmental stress cracking, leading to premature tensile failure. Biological attack may be regarded as chemical attack, as the microorganisms attracted to the geotextile initiate enzyme attack which involves a chemical process. Similarly, photochemical and irradiation chemical attack, although initiated by light, ultimately involve chemical degradation of the molecular structure of the polymer of the geotextile.

League tables have been published giving the weaknesses and strengths of various pure polymers subject to various modes of attack. These commonly state that polyester has good resistance to ultra-violet attack while the polyolefins have poor resistance. Although true, this and many of the other implications of such assessments are misleading in their generalisations. In the same way as the corrosion resistance of steel can be enhanced by the addition of nickel and chromium to produce stainless steel, the ultra-violet resistance of the polyolefins can be improved by the addition of carbon black and other stabilisers. In short, the resistance of a particular geotextile to a particular form of attack is not only related to geotextile structure and base polymer, but it is

Properties and their measurement

affected by the additives incorporated, and, in some instances, by the manufacturing process. What is still little understood is the reaction of geotextiles to complex environments. A geotextile may be resistant to ultra-violet attack alone or to sea water attack alone, and yet may degrade rapidly if subjected simultaneously to both elements.

Site quality control testing

The determination of geotextile properties for design purposes may involve highly sophisticated test techniques. Clearly, such test facilities would not be available even on large sites. In spite of this, there is still the need to ensure that the specified geotextile is indeed the actual geotextile delivered to site. To simplify test methods for quality control and to allow a meaningful comparison between different geotextiles, the notion of *index testing* has been introduced. Index testing involves the use of very simple techniques, which do not give definitive design parameters for a geotextile, but do give reproducible results, suitable for quality control and comparison of geotextiles.

As a bare minimum on any site, a check should be made that the correct geotextile is being delivered. This task will be eased by the adoption of a standard system of identification. To identify each roll or package of geotextile, the following information might be provided: manufacturer's name, commercial name of geotextile, method of manufacture and constituent materials, mass per unit area, nominal thickness, and dimensions and weight of geotextile in roll.

On larger sites with a weigh bridge facility, the delivered weight may be checked and compared with the product of mass per unit area and delivered area. Even if this facility is not available, a simple check on the mass per unit area may be made using basic equipment such as a balance or scales. A specific area of material should be accurately cut out and bound into a bundle that may conveniently sit upon the weighing device. With due allowance for the weight of the binding, the mass per unit area of the fabric may be calculated to at least within 10 g/m^2. This should indicate clearly whether the correct grade of fabric has been delivered.

Over and above this simple check, the frequency and degree of quality control testing will tend to be a function of the application and the risk involved in that application. High risk applications are clearly those in which the consequences of geotextile failure would be very serious and extensive, such as the use of geotextile filters in dams and geotextiles as soil reinforcement. In these instances, testing of every roll, or at least every other roll, would not be inappropriate. In such demanding applications, the determination of the most important property (pore size for filters, tensile strength for reinforcement) should be made in addition to the more basic quality control mentioned above. Ideally, testing should be performed on site, which would clearly necessitate the establishment of a small laboratory. If other materials testing were being performed on site this should not cause a problem; otherwise, it may be more economical to transport samples regularly to a central testing laboratory.

For 'fail safe', or low risk, applications, such as the use of geotextiles as separators in unpaved roads, only the basic index tests need be carried out. In addition, the frequency of sampling may be relaxed to testing a sample from every one in ten or twenty rolls.

4
General construction techniques

Construction techniques vary from application to application, although certain techniques, such as site reception and storage, are common to most applications. Whatever the application the objective is the same, namely to place the correct geotextile in the correct location without impairing its properties during the construction process. In essence, working with geotextiles is simple, but like any other undertaking, the ease of operation is a function of how carefully the execution of the works has been planned, taking into account prevailing site conditions.

Site reception

Geotextiles are dispatched to site in rolls wound on to robust cylindrical cardboard or plastic formers, and in most cases are delivered by road. On arrival at site, a check should be made that the quantity being delivered is correct and that the geotextile is of the specified type. Only a cursory check can be made before the consignment is unloaded, as it is often difficult to gain access to rolls at the base of the stack. As mentioned, each roll should be labelled by the manufacturer to give basic information on the product. When delivered, all the rolls should be wrapped in a protective layer of plastic, in which case selected rolls will have to be opened to reveal the geotextile. This involves slitting the wrapper, which should be pulled away from the geotextile before cutting to avoid damage to the geotextile inside. If the checker has previously been supplied with a hand sample of the specified geotextile, then a comparison can be made. On completion of inspection, the slit in the wrapper must be sealed with wide adhesive tape, as the wrapper is the only protection accorded to the geotextile. Before delivery is made, thought should be given to the location of storage areas. With due regard to access and safety,

Geotextiles handbook

storage areas should be located as close as possible to the point of end use, in order to minimise subsequent handling and transportation; therefore, in the case of a road project it may be useful to have several storage areas positioned along the site.

Storage

The objective in storing geotextiles is to prevent damage to, or deterioration of, the product before use. As the geotextile is generally delivered with a protective outer wrapper, it is usually adequate to stack the rolls directly on the ground, provided that this is even, well-drained and free from sharp projections such as rocks, stumps of trees or bushes. Instructions should be given regarding the setting aside of any rolls to be used to take samples for quality control. A prime cause of environmental degradation is sunlight. However, provided that the protective wrapper is intact, this poses no problem unless the product is to be stored for long periods of time in extreme climates; in this case, some form of shading is required, unless the wrapper is of opaque material, to give protection against ultra-violet light attack. Where the product is to be used in long-term reinforcing applications, extra care should be taken to guard against ultra-violet attack, as the long-term effects of short-term exposure to ultra-violet light are not well understood.

It cannot be emphasised too strongly that great care should be taken not to damage the protective wrapping. If damage occurs and the wrapping is beyond repair, the roll should be stored to prevent ingress of water. Without this, the geotextile, particularly the non-wovens, will absorb water, thereby causing the weight of the roll to increase, possibly to an unmanageable level. Further complications arise if damp rolls are allowed to freeze, as this makes unrolling difficult, if not impossible. Where geotextiles are to be used as filters, it is important to keep the wrapper intact to give protection against ingress of dust or mud. Should contamination or damage occur, the first few turns, or more, of the fabric should be discarded until the unaffected sections of the roll appear.

Handling

Handling methods are very much affected by the weight and size of the geotextile roll. Roll weights vary between approximately

General construction techniques

25 kg for short narrow rolls of geogrid to something in excess of 1000 kg for the long wide rolls of heavyweight fabric. More typically, roll weights vary between 75 kg and 150 kg. The lighter weight rolls can be manhandled, whereas any lifting or cross-site transportation of the heavier rolls requires mechanical assistance. Fig. 28 shows a purpose made frame for attachment to a front loader. The frame, because it is rigid, allows complete control of the roll and accurate placing. An alternative method of handling is shown in Fig. 29. This method involves the use of a robust metal spindle which is passed through the hollow former on which the geotextile is wound. The spindle projects from each end of the roll to allow the connection of a lifting sling to each end of the spindle.

Fig. 28. (Courtesy ICI Fibres)

Fig. 29. *(Courtesy MMG Civil Engineering Systems)*

General construction techniques

The sling is then connected to the boom of any convenient excavator to allow lifting. This technique requires two site operatives to steady the roll and to assist in accurate placement. Whatever lifting method is used, the load should be taken through the roll former and not directly by the geotextile. What should be avoided is lifting with chain or wire slings wrapped directly round the roll, as this practice is likely to damage the geotextile.

Laying

Many applications, such as separation between the formation and sub-base of a road, require geotextiles to be laid over large areas. Where the ground is comparatively dry and firm, this presents few problems; however, a modicum of site preparation is required where the geotextile is to be laid on a 'green fields' site. This involves the removal of sharp objects, such as rocks, stumps of trees or bushes, which might puncture or tear the fabric. In addition, any significant hollows or unevenness in the site should be filled. Without this precaution, the less extensible geotextiles tend to span such hollows and, subsequently, may tear when they are loaded. Obviously, these problems do not arise when the original ground level has been graded to some predetermined formation level. Following site preparation, the geotextile is rolled out into position, allowing an overlap between adjacent sheets (Fig. 30). To maintain the position of the geotextile before covering with sub-base or other fill, the edges of the sheets must be weighted. This is best achieved by placing shovelfuls of fill along the overlaps at approximately one metre intervals (Fig. 31). If the site is particularly exposed to wind, this weighting process needs to be extended to weighting the body of the sheet to prevent billowing or uplift. This problem can be minimised by not laying the geotextile too far in advance of placing the sub-base or other fill which is finally to cover the geotextile.

One advantage in using geotextiles is that they allow construction over poor ground. As ground conditions become worse, the practice of rolling out the geotextile becomes more difficult, and other methods have to be adopted. This generally entails unrolling the geotextile on firm dry ground and then manhandling it to place in position (Fig. 32). Where necessary, the geotextiles are joined by sewing or some other mechanical means before placing. When large quantities of fabric have to be placed over poor ground, it

Geotextiles handbook

Fig. 30. (Courtesy Amoco Fabrics (UK) Limited)

General construction techniques

Fig. 31. (Courtesy Don and Low Limited)

Fig. 32. (Courtesy Hoechst UK Limited)

might prove more economical to consider some form of mechanical assistance; however, even these techniques still prove labour intensive.

Jointing

In the majority of applications, it is adequate to form a simple unbonded lap joint. The overlap used is generally a minimum of 300 mm, but in applications where the geotextile is subject to tensile stress, the overlap must be increased or the sheets of geotextile bonded. Without bonding or sufficient overlap of joints, problems will occur when placing fill over geotextiles laid on weak bearing soils. As the fill advances along the geotextile, the fill and the geotextile can move down appreciable distances. If, under these conditions, only a small lap length is employed, the unbonded joint may pull apart, allowing fill adjacent to the joint to penetrate the weak bearing soil. Once this has been allowed to happen, it is difficult to recover the situation without causing damage to the fabric. Before resorting to the use of bonded joints, it is advisable to consider larger overlaps, as lap lengths of up to one metre are generally more economical than bonded joints. The expedient of using increased lap lengths is generally unacceptable if the geotextile has to resist high sustained tensile stress, as in the case of a basal reinforcing layer to an embankment over soft clay.

There are several methods for bonding joints, including welding and gluing. However, the only practical methods for on-site use are sewing and stapling. Both these methods can be applied to woven and non-woven fabrics, while geogrids can only be 'sewn' using a robust cord threaded through the grid apertures. Sewn seams can be formed in a variety of ways, with two of the commonest forms shown in Fig. 33. The simplest seam (Fig. 33(a)) is made by folding back approximately 100 mm of fabric along the edge of each geotextile and placing these upstands face-to-face before sewing. The stitching, which passes through two thicknesses of fabric, should be at least 50 mm back from the free edges of the fabric. For obvious reasons, this type of seam is often called a 'prayer seam'. In theory, a more robust seam (Fig. 33(b)) is the lapped or J-seam, where the stitching has to penetrate four thicknesses of fabric; however, experience shows that slippage of the internal length of fabric out of the seam occurs, thus reducing its strength.

General construction techniques

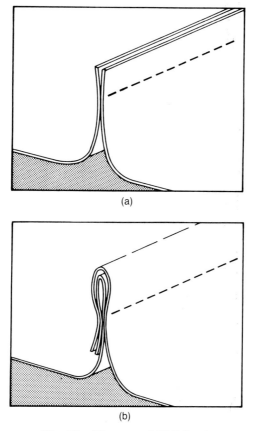

Fig. 33. (Courtesy ICI Fibres)

Ease of seaming on site is very much a function of ground conditions. Where these hamper handling of the fabric, the joint may be made by laying the sheets side by side and lifting the fabric off the ground to sew (Fig. 34). Once the seam has been made, the fabric is pulled as taut as possible, leaving the sewn seam standing upright and ready for inspection.

An alternative method of positioning and jointing the fabric, which in practice is far easier to perform, involves the four stages shown in Fig. 35: in stage 1, the first sheet is unrolled; in stage 2, the second sheet is laid directly on top of the first; stage 3 involves sewing the two sheets together along one edge using a prayer

Geotextiles handbook

Fig. 34. *(Courtesy Amoco Fabrics (UK) Limited)*

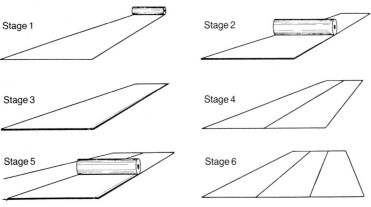

Fig. 35. *(Courtesy Rhône-Poulenc Fibres)*

General construction techniques

Fig. 36. (Courtesy UCO)

seam; and in the final stage, the top sheet is unfolded to leave the pair of sewn sheets side by side. This process can be repeated as required—for example, by rolling a third sheet on top of the unfolded second sheet, sewing and then unfolding as shown in stages 5 and 6 of Fig. 35. Comparison of Figs 34 and 36 shows the latter approach to be more orderly, provided that ground conditions are dry. With this technique the sewn joint must be inspected immediately, because when the sheets are unfolded the seam is on the underside and therefore not exposed for inspection.

Sewing is carried out using a portable bag closing machine operated by mains/battery or air. Two types of stitch may be used: the double thread lock stitch or the single thread lock stitch. The former has the advantage that it is less prone to unravel should one of the stitches be cut. For most non-wovens and lightweight wovens, it is possible to obtain seam strengths equal to some 50% of the fabric strength; however, efficiency of the seam will be governed by the strength of the thread. As the strength of the fabric increases, seam efficiency may drop off. In applications where seam strengths are critical, it is advisable to test seams using uniaxial tensile or tear testing similar to that employed for intact fabrics. Although sewing is the most widely used method of making a bonded joint, stapling often provides a useful alternative. This technique should be used only in association with the lapped or J-seam, using corrosion resistant staples applied by a heavy duty industrial stapling machine.

Cutting

Cutting of geotextiles on site is a labour-intensive, time-consuming operation, which in most cases can be avoided by forward planning. The maximum geotextile width is generally 5·3 m; however, many manufacturers produce the whole, or the more popular part, of their product range in lesser widths of approximately 2·0, 3·5 or 4·5 m, thereby obviating the need to cut a geotextile along its length to produce a narrower width. Although the total width of an area to be covered (allowing for overlaps or joints) will rarely be an exact multiple of available widths, there is less wastage of time and money if slightly larger overlaps are allowed to take up the excess width, than if the geotextile is cut on site.

If site cutting should prove unavoidable, there are various methods from which to choose, the simplest being that of using a sharp knife or shears to cut the fabric once it has been rolled out. For heat bonded non-wovens requiring longitudinal cutting to give a uniform reduction in width, it is easier to cut the fabric while it is still rolled up. This can be done using a chainsaw, or a hand hacksaw. A powered saw should not be forced through the roll too rapidly because the heat generated may fuse the fabric at the cut end, thus making unrolling difficult. This technique must not be used with woven fabrics as the yarns will foul the saw mechanism. For fabrics that prove difficult to cut using these techniques, consideration should be given to the use of a portable 'hot-knife' cutter, which locally melts through the fabric while at the same time fusing loose ends which might otherwise fray. This technique should, of course, only be used after unrolling.

One application in which cutting and tailoring of the geotex-

General construction techniques

tile may be necessary is the lining of hazardous waste lagoons before placement of an impermeable geomembrane liner. In this instance, it is vital to produce a detailed laying plan which might be used as a basis to prefabricate difficult joints or shapes.

Site damage and repair

Damage repair, like on-site cutting, can best be avoided by forward planning and, for larger projects, by field trials. The majority of visible site damage is in the form of puncturing, bursting or tearing. All of these can be minimised by taking reasonable precautions in site preparation, as described, and reasonable care in construction. A more insidious form of damage is that caused by the environment. Prime among these is ultra-violet exposure, which can cause dramatic decreases in strength without visible change in the fabric. This type of damage can usually be avoided by not laying more geotextile in a day than can be covered by fill in that same day. This is particularly true in marine works, such as revetments, where exposure to the combined effects of sunlight and sea water can cause certain geotextiles to disintegrate in a matter of weeks. Even more significant can be the effects of wave action on unprotected non-woven geotextiles, in disassociating the filament mass in as little as twenty-four hours. Other environments to be avoided are those created by spillage of fuel oils and petrol.

More evident damage is caused by careless construction techniques. Geotextiles must always be laid so that they are evenly supported by the ground, even if this is weak. This requirement applies equally to vertical applications, such as the sides of drainage trenches, where the geotextile must follow the contours of the trench or be allowed enough slack to achieve this requirement as filling of the trench progresses. Once the geotextile is laid, it should not be trafficked until an adequate layer of fill is placed over it, thus affording some protection. The one exception to this rule is where a heavyweight geotextile is used, which is specifically designed to be directly trafficked by vehicles. By the same token, the blades or buckets of construction plant must not be allowed to come into contact with the fabric during filling operations. While the contractor is not expected to perform with the precision of a watchmaker, practices such as those depicted in Fig. 37 should not be permitted if the geotextile is expected to fulfil its design function.

Geotextiles handbook

Fig. 37. (Courtesy Chemie Linz AG)

Inevitably, damage will occur at some time and remedial measures will have to be considered. These will tend to reflect the importance of the application and the consequences of failure should this occur. For the more critical structures, such as reinforced soil walls and embankments, it is safest to remove the damaged section of geotextile entirely and replace it with undamaged material. In this application, a certain degree of damage may be acceptable, provided that this has been allowed for at the design stage. With many structures, such as armoured revetments, the geotextile will play a vital part in long-term stability, yet constitutes only a small proportion of the overall project cost. Therefore, again it makes sense to replace damaged sections.

In lower risk applications, where the fabric is not subject to significant tensile stress or dynamic water loading, it is permissible to patch the damaged area. This is best achieved by removing the

General construction techniques

Fig. 38. (Courtesy Don and Low Limited)

fill, by hand, to expose the damaged geotextile and an area of undamaged geotextile at least 300 mm wide around the damaged area. A piece of undamaged fabric is then cut to a size large enough to cover the exposed geotextile. This patch is carefully bedded over this area (Fig. 38), covered with fill and compacted in the normal manner.

5
Applications and construction techniques

It is beyond the scope of this small book to cover specific construction techniques in detail, as these tend to vary from site to site and from application to application. However, the following describes some common applications and also sets out rudimentary construction techniques.

Temporary roads
Temporary roads are frequently required to give site access across weak bearing soils such as soft clays, water bearing silts or peats. Even with the stronger clay formations, trafficking can become difficult during and after wet weather. The problem can be overcome by placing a carpet of granular material—such as sub-base, crushed rock or hardcore—over the weak formation to a sufficient depth to spread vehicle wheel loads to such an extent that they do not overstress the weak formation. Despite this precaution, it is frequently found that extensive ruts form in the surface of such an unpaved road along the paths of the vehicles' wheels. These surface ruts are reflected at depth in the surface of the weak formation where the granular fill tends to punch into the soil and become ineffective. This loss of granular material causes an effective reduction in the original 'carpet' thickness and a consequential increase in stress to the formation. This, in turn, leads to further local loss of 'carpet' and a vicious circle sets in of increased rutting. Eventually, the temporary road either becomes impassable or requires constant maintenance. These problems can be minimised by incorporating a suitable geotextile or geogrid in the road construction.

Before the granular sub-base is placed, the formation width is covered with geotextile (Fig. 39). Although brick hardcore may be used, large pieces of intact brickwork are totally unsuitable. Once the granular material is placed over the geotextile, two benefits result. Firstly, the geotextile acts as a separator and prevents mixing of the granular material and weak underlying soil; this allows the original thickness of the granular carpet to be maintained, as the geotextile prevents loss of granular material by punching. Secondly, through absorbing horizontal shear stress, the geotextile reduces the load transmitted to the formation, which may allow a reduction in the thickness of the granular layer otherwise needed without a geotextile. To mobilise this effect, there must be a limited amount of rutting to cause a deflection in the geotextile.

A further requirement is that the geotextile should be anchored on each side of the road. The bond length to give this anchorage varies according to the nature of the formation soil, the granular fill and the geotextile, and typically is around 1·0–1·5 m. This can be achieved by extending the geotextile beyond the required running width of the road (marked by sleeper kerbs or similar, as illustrated in Fig. 40(a)) or by providing an equivalent bond length by burying the fabric in shallow trenches (Fig. 40(b)) or by wrap around (Fig. 40(c)). The techniques for placing the geotextile are similar to those used for installing a fabric at the sub-grade/sub-base interface of a permanent road, which are described in the next section.

Permanent roads

The lower layers in a permanent road pavement generally comprise granular materials which are used to form the sub-base. If the sub-grade is a low CBR cohesive soil, it may be necessary to employ a capping layer to give a sound formation on which to place and compact the sub-base. Although the road may be constructed successfully, the long-term interaction between a cohesive sub-grade and the unbound granular layers placed on it, can have a dramatic effect on the design life of the entire pavement.

Road pavements are rarely impermeable: water can enter the road construction and come finally to rest near the surface of the sub-grade where it can cause softening of the soil. When this water is put under pressure by dynamic traffic loading, high water pressures are induced that can jet or pump the softened formation soil into the granular layers above. As well as reducing the support

Geotextiles handbook

Fig. 39. (Courtesy Netlon Limited)

Applications and construction techniques

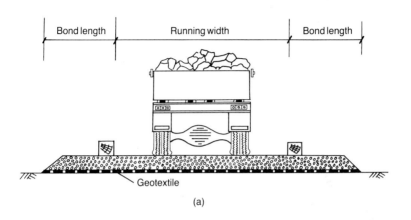

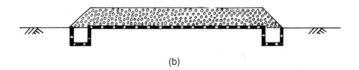

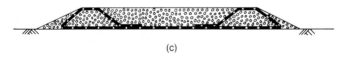

Fig. 40. Use of geotextiles in temporary road construction

Geotextiles handbook

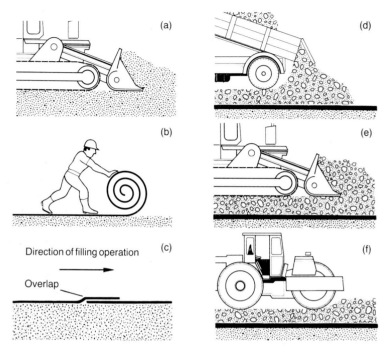

Fig. 41. *(Courtesy Rhône-Poulenc Fibres)*

offered to the sub-base, this can lead to a reduction in permeability and dynamic deformation modulus of the unbound layers, leading to premature failure of the pavement. Such problems can be ameliorated by the use of a suitable geotextile which, when installed at the sub-grade/sub-base interface, will tend to inhibit pumping.

The construction technique is summarised pictorially in Fig. 41, which shows six stages in the operation. Firstly, ground level is reduced to formation level; where the formation is susceptible to rutting, only tracked vehicles should be employed in the final strip of soil down to formation level (Fig. 41(a)). The geotextile is rolled into position (Fig. 41(b)), a lap joint being used at the end of each roll length (Fig. 41(c)). Sub-base or capping layer material is then placed over the geotextile by end tipping (Fig. 41(d)). Under no circumstances should construction plant be allowed to traffic directly on the geotextile. Next, the sub-base is spread and

Applications and construction techniques

levelled ready for compaction (Fig. 41(e)). At this stage and at the end tipping stage, the pile of sub-base material should not be allowed to become so high that it causes bearing capacity failure and heaving in the formation. Once the material has been spread and levelled, it is compacted (Fig. 41(f)) and should then be protected by construction of the bound pavement layers as soon as possible.

Repair of permanent roads

With time, the riding quality of a road surface will degrade through the formation of ruts and cracks which emanate from deep within the road construction. If the pavement is only surface-dressed or resurfaced, these deep seated cracks will again emerge to degrade the resurfacing. Although the effects of such 'reflective' cracking cannot be removed without reconstruction of the entire bound pavement depth, the use of a geotextile can serve to extend the service life of the resurfacing. The geotextile is installed between the old and the new surfacing to improve crack resistance. In installing the geotextile, it is vital that a good bond be achieved between old and new works. This involves first coating the old surfacing with a hot tar spray, or an emulsion, before rolling out the geotextile (Fig. 42). Once this is in place, the new surfacing is placed in the normal manner. Care must be taken to ensure that the laying temperature of the bituminous surfacing is well below that of the melting point of the geotextile.

Railway tracks

The sleepers of a railway track are supported by a layer of ballast which spreads the load into the formation soil. If the surface of the formation soil is fine grained or cohesive, then this can wet-up and be pumped under dynamic wheel loading in a similar way to the formation of a road. As can be seen from Fig. 43, this depletes the performance of the ballast, which has to be replaced or decontaminated. When a new track is to be laid over problem soils, or an old track is to be refurbished, the geotextile is placed on the formation before covering with clean ballast (Fig. 44). On many railways, sophisticated machinery has been developed that will lift the track, remove contaminated ballast, lay a geotextile and finally reinstate fresh, or decontaminated, ballast.

Geotextiles handbook

Fig. 42. (Courtesy Don and Low Limited)

Fig. 43. (Courtesy Don and Low Limited)

Applications and construction techniques

Fig. 44. *(Courtesy ICI Fibres)*

Embankments on soft ground

When embankments are constructed over weak soils such as soft clays, there can be problems with short-term instability in the form of deep-seated rotational slipping or transverse spreading of the embankment. Before the advent of geotextiles, these problems were overcome by building the embankment with very flat side slopes, or berms (Fig. 45(a)); in extreme cases, embankments have been constructed on piled foundations.

A much more economic solution can be achieved by using a basal layer of geotextile reinforcement, placed over the original formation before placing of embankment fill (Fig. 45(b)). If correctly designed and installed, the geotextile will impart tensile strength to the base of the fill, thereby resisting lateral spreading, rotational failure or extrusion of the underlying soft ground. The tensile restoring force required is usually across the width of the embankment and is likely to be very large (perhaps as much as 1000 kN/m). For these large loads, use is made of woven fabrics, or perhaps strips of strong material mounted on a weaker carrier fabric. In either case, it must be remembered that most reinforcing fabrics have orthotropic strength, which means they are strong in one direction but not so strong at right angles to this direction.

It follows then that it is vital to lay the geotextile in such a way that its axis of greatest strength coincides with the direction of maximum force applied by the embankment. With most reinforcing fabrics, the greatest strength is in the machine or lengthwise direction. Having the maximum strength in the machine direction

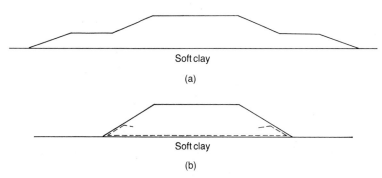

Fig. 45. *Embankments over soft ground: (a) with berms; (b) with geotextile reinforcement*

Applications and construction techniques

(or warp) is advantageous, as continuous lengths can be laid across the entire width of the embankment, without the need for joints to be made. Naturally, adjacent rolls are joined by sewing, but induced stresses at right angles to the machine direction (along the axis of the embankment) are much less.

Difficulty of filling over the geotextile increases as the foundation soil becomes weaker. For very soft formations, it is necessary to use special low bearing pressure tracked vehicles for spreading the fill. Once the geotextile has been laid, it is first necessary to anchor the two remote ends by constructing a bund along each toe and embedding the free ends of the fabric (Fig. 46(a)). The height of these bunds has to be judged by experience, but on soft ground it would not be more than 500–600 mm. Once the bunds are complete, the area between them can be filled to form a working platform (Fig. 46(b)). Before further filling, the embankment is advanced at the level of the working platform. This is likely to produce a bow wave of mud, which advances in the direction of construction and can impose high tensile stresses on the geotextile. The geotextile should not be laid too far ahead of the bow wave, as the drag forces will tear it; however, it should not be filled too near its free edge, or else this will be forced beneath the fill and lost.

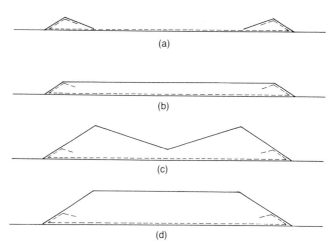

Fig. 46. *Stages in construction of basal geotextile reinforced embankment*

Geotextiles handbook

Once the working platform has been constructed, the main bulk of the fill can be placed in the sequences illustrated in Fig. 46(c) and (d). Placing of fill—working from the edges towards the centre of the embankment—serves to keep a more uniform tension in the geotextile and reduces the chances of the toes of the embankment being drawn together (as might occur if a large thickness of fill were placed in the centre of the embankment without the toes being weighted).

Construction of embankments over soft foundations calls for support from the geotextile reinforcement for a comparatively short period of time, extending to maybe a few years. Once the foundation soil has consolidated and gained strength, the geotextile should become redundant; thus no great demands are made on the long-term durability of the geotextile.

Walls and steep sided embankments

Unlike embankments on soft ground, walls and steep sided embankments need support from the geotextile or geogrid for their entire design life. Therefore, durability of the reinforcement is very important. The essence of construction is very simple, comprising the placing of selected fill incorporating horizontal layers of geotextile or geogrid reinforcement. As one lift of fill is completed, the reinforcement is rolled out over the surface of the fill, ensuring that it runs far enough back from the face to ensure an adequate bond length, as determined by the properties of both reinforcement and fill. Attention must be given to preventing spillage of fill from the face of the wall. This can be achieved by using 'hard' facing units tied to the reinforcement, or, if a 'soft' finish is desired, it is possible to encapsulate the next layer of fill with the geotextile reinforcement. In this case, a free length of reinforcement is left at the face sufficient to wrap around the next layer and to extend back over it to ensure adequate anchorage.

Slopes that are to be covered with vegetation can be constructed without any kind of formwork to support the reinforcement facing during filling. However, this does result in an irregular face (Fig. 47). A far better alignment and finish can be obtained by constructing a scaffold formwork (Fig. 48) against which the geotextile or geogrid can be supported during placement of fill. Formwork is generally necessary for constructing planar slopes steeper than 45°. This results in a far more pleasing finish (Fig. 49).

Applications and construction techniques

Fig. 47. (Courtesy UCO)

Geotextiles handbook

Fig. 48. (Courtesy British Aerospace plc)

Applications and construction techniques

Fig. 49. (Courtesy British Aerospace plc)

Geotextiles handbook

Drainage applications

Aggregate filled trench drains have been used to drain waterlogged land for many years. These tend to be installed at depths of typically one metre, although they can be installed at much greater depths if there is a need to depress the groundwater table significantly. In fine grained soils, the flow of groundwater towards the drain can carry soil particles. To prevent these particles from being carried into the aggregate fill of the drain, the grading of the fill has to be selected to act as a filter to the soil to be drained. This can prove to be very expensive.

With the availability of geotextiles with a wide range of pore sizes and permeabilities, it is a comparatively simple matter to select a fabric that will filter the soil to be drained. In this case, the drainage trench is simply lined with the filter fabric and backfilled

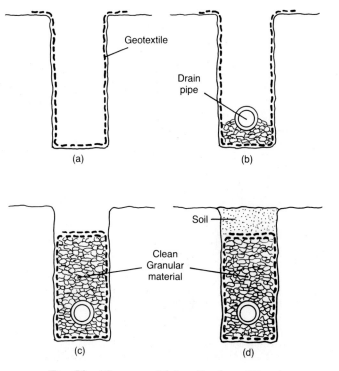

Fig. 50. (Courtesy Rhône-Poulenc Fibres)

Applications and construction techniques

with an aggregate that is coarse enough to act as a drainage medium. Where required, the trench drain can be constructed with a carrier pipe.

The construction sequence is shown diagrammatically in Fig. 50. First, the drainage trench is dug and lined with fabric (Fig. 50(a)). Before lining the trench, all sharp stones and other projections must be removed. Where the free edges of the fabric are laid on the ground at the edges of the trench, they should be held by stones or small piles of aggregate (Fig. 51). Next, the granular fill or drainage medium is laid on the bottom of the trench to act as bedding for any drainage pipe to be installed (Fig. 50(b)). Aggregate is then carefully placed to within 300 mm of the top of the drain. During this filling process, no attempt should be made to restrain the top of the fabric, as this is likely to be pulled downwards into the trench as the fill forces the fabric to follow the contours of the trench walls. Once filling is completed to this level, the free lengths of fabric are wrapped over the top of the aggregate (Fig. 50(c)). If the drain is to be used only for groundwater

Fig. 51. (Courtesy Don and Low Limited)

Geotextiles handbook

control, the top of the trench is sealed off with topsoil. If, conversely, the drain is expected to collect storm water run-off, the upper section of the drain should be filled with coarse aggregate. This, and the top of the fabric, will require some maintenance if the storm water to be collected carries significant quantities of soil fines.

Much the same effect can be achieved by using prefabricated fin drains such as those shown in Figs 52 and 53. These comprise a vertical water-conducting core, sandwiched between outer layers of woven (Fig. 52) or non-woven (Fig. 53) filter fabric. Fin drains are installed in an open trench (such as shown in Fig. 51) but can often be back-filled with the arisings from the trench rather than with expensive aggregate. Fin drains are particularly simple to use as drainage behind retaining walls or abutments, where they obviate the need for porous concrete, block work or aggregate drains, which are expensive to construct.

Fig. 52. (Courtesy Don and Low Limited)

Applications and construction techniques

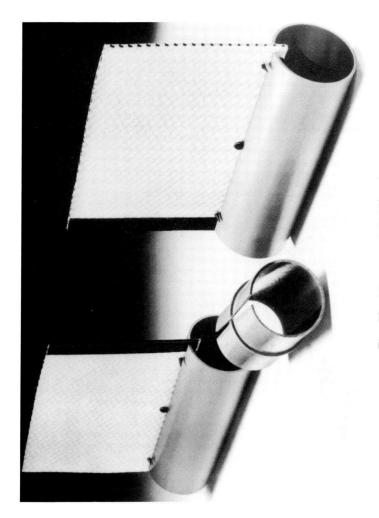

Fig. 53. (Courtesy ICI Fibres)

Geotextiles handbook

Fig. 54. (Courtesy BSP Limited)

Applications and construction techniques

Fig. 55. (Courtesy MMG Civil Engineering Systems)

71

Revetments

Geotextiles are applied extensively to erosion control problems, where they serve to prevent scour or pumping of erosion prone soils. Used in revetments to the banks of canals and other waterways, they are applied directly to the surface of the bank soil before being covered by a protective layer of granular fill and an outer layer of rip-rap or primary armour stone.

For less exposed applications, a much lighter armour stone can be used and may be applied directly to the surface of the geotextile (Fig. 54). Revetments, and erosion control in general, currently tend to be subject to a glut of proprietary systems. Typical of these are the so-called flexible armoured revetments, which consist of open precast concrete blocks strung on to longitudinal and transverse wires (or cords), to form a 'chain mail' unit which can be draped immediately on to the bank requiring protection. Such units are always underlain by a geotextile, which controls loss of fines from the bank soil. For large drainage channels or navigable inland waterways, where craft-induced erosion is not so severe, protection of the banks (and if necessary the beds) can be afforded by the use of thick webs or mats which can be faced on one side with a geotextile filter (Fig. 55).

6
Bibliography

1. *Proc. Int. Conf. on Flexible Armoured Revetments incorporating Geotextiles*, London, 1984. Thomas Telford, London, 1985.
2. *Proc. 2nd Int. Conf. on Geotextiles*, Las Vegas, 1982. Industrial Fabrics Association International, St Paul, Minnesota, 1982.
3. *Proc. 3rd Int. Conf. on Geotextiles*, Vienna, 1986. Österreichischer Ingenieur und Architektenverein, Vienna, 1986.
4. *Proc. Int. Conf. Geotextile Technology '84*, London, 1984. Construction Industry International, London, 1984.
5. *Proc. Conf. on Polymer Grid Reinforcement*, London, 1984. Thomas Telford, London, 1985.
6. *Proc. Int. Conf. on Use of Fabrics in Geotechnics*, Paris, 1977. Association Amicale des Ingenieurs Anciens Eleves de l'Ecole Nationale des Ponts et Chaussees, Paris, 1977.
7. *Geotextiles engineering manual*. Federal Highways Administration, National Highways Institute, Washington, 1985.
8. Giroud J. P. *Geotextiles and Geomembranes: definitions, properties and design*. Industrial Fabrics Association International, St Paul, Minnesota, 1984.
9. ICE Ground Engineering Committeee. *Model procedures and specifications for geotextiles*. Thomas Telford, London (in preparation).
10. Koerner R. M. *Designing with geosynthetics*. Industrial Fabrics Association International, St Paul, Minnesota, 1986.
11. Koerner R. M. and Welsh J. P. *Construction and geotechnical engineering using synthetic fabrics*. John Wiley, New York, 1980.
12. Rankilor P. R. *Membranes in ground engineering*. John Wiley, Chichester, 1981.

Geotextiles handbook

13. Van Zanten R. V. *Geotextiles and geomembranes in civil engineering*. A. A. Balkema, Rotterdam, 1986.
14. *Geotechnical fabrics report*. Published bimonthly by Industrial Fabrics Association International, St Paul, Minnesota.
15. *Int. J. Geotextiles and Geomembranes*. Published bimonthly by Elsevier Applied Science Publishers, Barking.

7
Compendium of product data

List of manufacturers

Amoco Fabrics
BTR Landscaper
Chemie Linz AG (now Polyfelt GmbH)
Don and Low Limited
Du Pont de Nemours International SA
Enka Glanzstoff BV
Fibertex APS
Godfreys of Dundee Limited
Hoechst AG
Huesker Synthetic GmbH
ICI Fibres Geotextile Group
LEC Limited
Malcolm, Ogilvie and Company Limited
Monsanto Europe SA/NV
Netlon Limited
Nicolon BV
Rhone-Poulenc Fibres
SA UCO NV

While every care has been taken in the compiling of this compendium, users are reminded that they should ascertain current product data for themselves direct from the geotextile manufacturer or supplier.

Amoco Geotextiles

the high performance range of Geotextiles from Amoco Fabrics

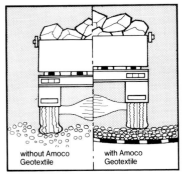

without Amoco Geotextile | with Amoco Geotextile

SEPARATION & REINFORCEMENT

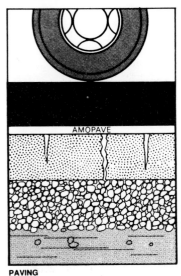

PAVING

AMOCO GEOTEXTILE

FILTRATION

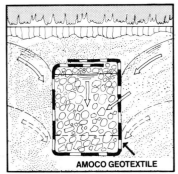

AMOCO GEOTEXTILE

DRAINAGE

Amoco Fabrics
A division of Amoco (U.K.) Ltd.
Suite 3D, No 1 Tabley Court, Victoria Street,
Altrincham, Cheshire WA14 1EZ.
Telephone 061-928 8616
Telex 668301

Amoco Fabrics
Niederlassung der Amoco Deutschland GmbH
Kostverlorenhof 2
NL-1183 HE Amstelveen
Telephone 020-43 89 01
Telex 14 577

Amoco Fabrics

Compendium of product data

Manufacturer
Amoco Fabrics
Niederlassung der Amoco Deutschland GmbH
Kostverlorenhof 2
1183 HE Amstelveen
The Netherlands
Telephone: 010 31 20 438901   Telex: 14577
Telefax: 010 31 20 459887

UK supplier
Amoco Fabrics
Suite 3 D
1 Tabley Court
Victoria Street
Altrincham
Cheshire
WA14 1E2
Telephone: 061 9288616   Telex: 668301
Telefax: 061 9281272

Product range
Amoco 6050, 6060, 6061, 6062, 6064, 6068, 6071
100% polypropylene woven tape fabric
In addition, Amoco Fabrics also produce composite fabrics.
On any of the above-mentioned styles, a non-woven fabric can be needled. The weight and type of the non-woven fabric depend on the requirements.
Scotlay 120, 200
Tape-on-tape slit film
100% polypropylene woven fabric
Amoco 4545
100% polypropylene non-woven fabric
Other non-woven products available on request
AmoPave 4599
100% polypropylene non-woven fabric
This fabric is specially designed for use as an asphalt overlay fabric. It will retard reflective cracking and create an impermeable membrane in the road.

		Test reference	Product reference: AMOCO GEOTEXTILE						
			6050	6060	6061	6062	6064	6068	6071
Physical characteristics	Mass per unit area (g/m^2)	DIN 53854	95	135	210	190	325	470	290
	Thickness (mm)	DIN 53855	0.3	0.4	0.8	0.5	1.3	1.6	1.3
	Pore size (µm)	Delft Hydraulics	210	150	265	145	180	260	350
Strength	50 mm strip tensile test	Max. load (N/50 mm)	DIN 53857	850/850	1200/1100	2300/1250	1600/1500	2400/2600	4000/3400	2300/2300
		Extension at max. load (%)		15/15	15/15	20/12	15/15	24/12	15/12	20/12
	Grab tensile test	Maximum load (N/25 mm)	DIN 53858	500/450	750/700	1300/700	1200/900	1400/1500	2300/1900	1300/1300
	Plane strain tensile test	Tensile strength (kN/m) Extension at max. load (%)								
Puncture and tear	CBR plunger test	Maximum load (N) ($\bar{x} - s$)	DIN 54307	2050	3300	5000	4800	7000	10500	6500
	Drop cone test	Hole width (mm)		17	14	—	8	7.5	—	—
	Tear strength	Maximum load (N)	ASTM D 2263	220/220	400/400	700/350	450/450	700/550	—	—
Permeability	Normal to the plane of the geotextile	Normal pressure = kPa	Flow rate (l/s/m^2)	Delft Hydraulics (10 cm head)	10	13	45	20	25	30	40
			k(m/s)		2×10^{-5}	5×10^{-5}	3×10^{-4}	2×10^{-4}	3×10^{-4}	6×10^{-4}	4×10^{-4}
		Normal pressure = 2 kPa	Flow rate (l/s/m^2)	Franzius-Institut, Hannover	2×10^{-5}	1×10^{-4}	—	1×10^{-4}	4×10^{-4}	6×10^{-4}	5×10^{-4}
			k(m/s)								
	Width(s) available × length (m)			5.2×200	3.5×200 5.2×150	5.2×100	5.2×100	5.2×100	5.05×100	5.2×100

Note: Where two values are given, they refer to the lengthwise/crosswise directions.

		Test reference	Product reference: SCOTTLAY	
			120	200
Physical characteristics	Mass per unit area (g/m²)		120	200
	Thickness (mm)		0.3	0.4
	Pore size (μm) O_{90}		250	210
Strength	50 mm strip tensile test	Max. load (N/50 mm)	800	1750
		Extension at max. load (%)	22/12	18/6
	Grab tensile test	Maximum load (N/25 mm)	ASTM D 1682 400	950
	Plane strain tensile test	Tensile strength (kN/m)	16	35
		Extension at max. load (%)	22/12	18/6
Puncture and tear	CBR plunger test	Maximum load (N)	2400	3300
	Drop cone test	Hole width (mm)	15	12
	Tear strength	Maximum load (N)	BS 4303 180	400
Permeability	Normal to the plane of the geotextile	Normal pressure = kPa, Flow rate (l/s/m²)	(10 cm head) 15	10
		k(m/s)	6×10^{-5}	4×10^{-5}
		Normal pressure = kPa, Flow rate (l/s/m²)		
		k(m/s)		
	Width(s) available × length (m)		4.5 × 100	4.5 × 100

Note: Where two values are given, they refer to the lengthwise/crosswise directions.

		Test reference	Product reference: AMOCO GEOTEXTILE (CEF)
			4545
Physical characteristics	Mass per unit area (g/m^2)		150
	Thickness (mm)		1·3
	Pore size (μm)	Delft Hydraulics	100
Strength	50 mm strip tensile test	Max. load (N/50 mm) Extension at max. load (%)	400/500 40
	Grab tensile test	Maximum load (N/25 mm)	450
	Plane strain tensile test	Tensile strength (kN/m) Extension at max. load (%)	8/10 22/30
Puncture and tear	CBR plunger test	Maximum load (N) ($\bar{x} - s$)	—
	Drop cone test	Hole width (mm)	—
	Tear strength	Maximum load (N) ASTM D 2263	200
Permeability	Normal to the plane of the geotextile	Flow rate (l/s/m^2) k(m/s) Normal pressure = kPa Flow rate (l/s/m^2) k(m/s) Normal pressure = kPa	100 5×10^{-3}
	Width(s) available × length (m)		4·6×100

Note: Where two values are given, they refer to the lengthwise/crosswise directions.

		Test reference	Product reference: AMOPAVE (Asphalt overlay fabric)
			4599
Physical characteristics	Mass per unit area (g/m²)		140
	Thickness (mm)		0·9
	Pore size (μm)		—
Strength	50 mm strip tensile test	Max. load (N/50 mm)	400
		Extension at max. load (%)	40
	Grab tensile test	Maximum load (N/25 mm)	450
	Plane strain tensile test	Tensile strength (kN/m)	
		Extension at max. load (%)	
Puncture and tear	CBR plunger test	Maximum load (N) ($\bar{x} - s$)	
	Drop cone test	Hole width (mm)	
	Tear strength	Maximum load (N)	—
Permeability	Normal to the plane of the geotextile	Normal pressure = kPa Flow rate (l/s/m²) k(m/s)	
		Normal pressure = kPa Flow rate (l/s/m²) k(m/s)	
	Width(s) available × length (m)		3·8×100

Note: Where two values are given, they refer to the lengthwise/crosswise directions.

Geotextiles handbook

Manufacturer

BTR Landscaper
P.O. Box 3
Centurion Way
Farington
Preston
Lancashire
PR5 2RE
Telephone: 0772 421711   Telex: 67530

UK Supplier

Details of the UK distribution network are available from the manufacturer.

Product range

Stripdrain is a composite drain comprising a deep-dimpled core with a non-woven heat bonded or needlepunched geotextile on both sides.
Core high density polyethylene, geotextile polypropylene/polyethylene

Cordrain is similar to Stripdrain, but the geotextile is on one side of the core only.

Hi-drain is a specific high strength version of *Stripdrain* for use as a highway sub-base drain.

Compendium of product data

Manufacturer
Chemie Linz AG (now Polyfelt GmbH)
St Peter-Strasse 25
A-4021 Linz
Austria
Telephone: 010 43 732 666 381   Telex: 02 1324
Telefax: 010 43 732 667859

UK supplier
Chemie Linz UK Limited
12 The Green
Richmond
Surrey
TW9 1PX
Telephone: 01 948 6966   Telex: 924941

Product range
Polyfelt TS 500, 600, 700, 750, 800 (standard grades)
       TS 420, 550, 650, 006, 008 (special grades)
Ecofelt 021T, 022T
Needlepunched continuous filament
100% polypropylene

Polyfelt® TS
High Technology Multifunction Geotextiles

Available in a range of grades designed to meet a variety of site specific conditions

Every Polyfelt grade has the following important properties:
- Guaranteed multidirectional tensile properties
- High puncture and tear resistance
- Elongation under initial load
- High permeability
- Defined filtration capabilities
- In-plane drainage of liquid and gasses
- High resistance to chemical solutions
- High UV-resistance
- Weldable jointing to eliminate overlap wastage
- Professional design and application advice based on the combination of theoretical research and practical evaluation.

The New Polyfelt® TS Design Manual

Chemie Linz have now produced technical manual for the design and construction of projects utilizing the mechanically bonded polypropylene nonwoven geotextile Polyfelt® TS.

Price:
Austrian Schilling
500.–

Typical Geotextile Applications and Principle Functions

Geotextile Functions / Areas of Application	Separation	Filtration	Drainage	Reinforcement	Protection	Sealing
Unpaved Roads	●	○	○	○		
Repaving				○		●
Railroads	●	●				
Hydraulic Construction	○	●				
Drainage	○	●	○			
Sports Fields	●	●				
Embankments	●	○	○	○		
Vertical Drains		○	●			
Retaining Walls			○	●		
Tunnels			●		●	
Geomembrane Containments			○	○	●	

● Primary Functions   ○ Secondary Functions

Precise design information for all applications of Polyfelt is available on request.

Polyfelt® TS
The key to international civil engineering solutions

CHEMIE LINZ
Chemie Linz AG
St.-Peter-Strasse 25, P.O. B
A-4021 Linz/Austri
Telephone (0732)
Telex 021

Product reference: POLYFELT (Standard grades)

Category	Property	Test reference	TS 500	TS 600	TS 700	TS 750	TS 800
Physical characteristics	Mass per unit area (g/m²)	AFNOR GO-104, DIN 53854	140	200	280	350	400
	Thickness (mm) 2 kPa / 100 kPa / 200 kPa	EDANA 30074, DIN 53855/3	1·50 / 0·70 / 0·60	2·00 / 0·95 / 0·75	2·60 / 1·30 / 1·00	3·00 / 1·55 / 1·25	3·30 / 1·80 / 1·45
	Pore size (μm) D_w	Franzius-Institut, Hannover	120	110	90	80	70
Strength	50 mm strip tensile test: Max. load (N/50 mm) / Extension at max. load (%)	DIN 53857/2, NFG 07-001, ASTM D 1682	430 / 50–80	600 / 50–80	860 / 50–80	1000 / 50–80	1100 / 50–80
	Grab tensile test: Maximum load (N/25 mm)	DIN 53858, NFG 07-120, ASTM D 1682	510	760	1080	1320	1500
	Plane strain tensile test: Tensile strength (kN/m) / Extension at max. load (%)						
Puncture and tear	CBR plunger test: Maximum load (N) ($\bar{x} - s$)	DIN 54307	1350	1800	2600	3000	3400
	Drop cone test: Hole width (mm)	Technical Research Centre of Finland	15·4	12·0	10·0	8·1	7·2
	Tear strength: Maximum load (N)	DIN 53363	180	275	365	420	500
Permeability	Normal to the plane of the geotextile — Normal Pressure = 2 kPa: Flow rate (l/s/m²) / k (m/s)	Franzius-Institut, Hannover	330 / 5×10^{-3}	250 / 5×10^{-3}	190 / 5×10^{-3}	130 / 4×10^{-3}	120 / 4×10^{-3}
	Normal pressure = 200 kPa: Flow rate (l/s/m²) / k (m/s)	(10 cm head)	118 / 6×10^{-4}	80 / 6×10^{-4}	58 / 6×10^{-4}	55 / 6×10^{-4}	52 / 6×10^{-4}
	Width(s) available × length (m)		2/4 × 250	2/4 × 175	2/4 × 125	2/4 × 100	1·9/3·8 × 100

Note: Where two values are given, they refer to the lengthwise/crosswise directions.

		Test reference	POLYFELT (Special Grades)					ECOFELT	
			TS 420	TS 550	TS 650	TS 006	TS 008	O21T	O22T
Physical characteristics	Mass per unit area (g/m²)	AFNOR G0-104 DIN 53854	130	180	235	500	700	90	110
	Thickness (mm) 2 kPa / 100 kPa / 200 kPa	EDANA 30074 DIN 53855/3	1·40 / 0·55 / 0·50	1·80 / 0·90 / 0·70	2·20 / 1·10 / 0·85	3·90 / 2·30 / 1·90	5·20 / 3·20 / 2·60	1·10 / 0·45 / 0·40	1·20 / 0·50 / 0·45
	Pore size (μm) D_w	Franzius-Institut, Hannover	120	110	100	60	60	130	130
Strength	50 mm strip tensile test: Max. load (N/50 mm) / Extension at max. load (%)	DIN 53857/2 NFG 07 001 ASTM D 1682	400 / 50/80	540 / 50/80	720 / 50/80	1200 / 80/120	1450 / 80/120	260 / 40/70	320 / 40/70
	Grab tensile test: Maximum load (N/25 mm)	DIN 53858 NFG 07 120 ASTM D 1682	480	650	900	—	—	315	400
	Plane strain tensile test: Tensile strength (kN/m) Extension at max. load (%)								
Puncture and tear	CBR plunger test: Maximum load (N) ($\bar{x} - s$)	DIN 54307	1200	1500	2200	2800	3800	900	1050
	Drop cone test	Technical Research Centre of Finland	16·0	13·0	11·0	—	—	25·4	21·5
	Tear strength: Maximum load (N)	DIN 53363	150	225	300	—	—	110	130
Permeability	Normal to the plane of the geotextile: Normal Pressure = 2 kPa Flow rate (l/s/m²) / k(m/s)	Franzius-Institut, Hannover	360 / 5×10^{-3}	280 / 5×10^{-3}	230 / 5×10^{-3}	100 / 4×10^{-3}	60 / 3×10^{-3}	450 / 5×10^{-3}	420 / 5×10^{-3}
	Normal pressure = 200 kPa Flow rate (l/s/m²) / k(m/s)		120 / 6×10^{-4}	80 / 6×10^{-4}	60 / 6×10^{-4}	30 / 6×10^{-4}	18 / 5×10^{-4}	160 / 6×10^{-4}	130 / 6×10^{-4}
	Width(s) available × length (m)		2/4 × 250	2/4 × 200	2/4 × 150	2·5/5 × 150/75	2·5/5 × 130/65	2/4 × 300	2/4 × 300

Note: Where two values are given, they refer to the lengthwise/crosswise directions.

Compendium of product data

Manufacturer
Don and Low Limited
Geotextiles Division
St James Road
Forfar
Angus
DD8 2AL
Telephone: 0307 65111   Telex: 76552 DON LOW G
Telefax: 0307 68474

UK Suppliers
Frank Parker & Co. Ltd
W.W. Hall Ltd

Various depots throughout the UK
Contact manufacturer for further information

Product range
Lotrak general purpose geotextiles
10/7, 13/12, 16/15, 35/30, 45/45
Plain weaves using extruded tapes and slit fibrillated film
100% polypropylene

Lotrak high permeability geotextiles
MT22/16, MT22/20
Plain weaves using monofilament warp and extruded tape weft
100% polypropylene

Lotrak heavyweight geotextiles
50/50, 85/85, 120/120, 200/40, 320/60
Woven fabrics using slit fibrillated tapes and yarns
100% polypropylene

Lotrak composites
200, 400, 400C, 600, 45/45/200, 45/45/600.
Lotrak 16/15 or Lotrak 45/45 with staple fibre fleece needled to one side
100% polypropylene

Lotrak Geotextiles

Don & Low

DON & LOW LIMITED, ST JAMES ROAD, FORFAR, DD8 2AL, SCOTLAND.
TEL 0307 65111 TELEX 76552 DONLOW G FAX 0307 68474

Trammel fin drain
F30, F40, F50, Cut off
Laminated filter drain incorporating extruded mesh core sandwiched between one of three grades of woven monofilament-on-tape filter fabric or impermeable membrane

Trevira spunbond (Product data given on page 107)
13/130, 13/150, 11/180, 11/200, 11/270, 11/300, 11/360, 11/500
Needlepunched continuous filament
100% polyester

	Test reference	Product reference: LOTRAK (GENERAL PURPOSE)				
		10/7	13/12	16/15	35/30	45/45
Physical characteristics						
Mass per unit area (g/m²)		90	100	120	205	240
Thickness (mm)		0.3	0.3	0.3	0.5	0.65
Pore size (μm) O_{50} / O_{90}		160 / 230	210 / 290	190 / 300	190 / 230	430 / 500
Strength						
50 mm strip tensile test — Max. load (N/50 mm) / Extension at max. load (%)	BS 4859	500/350 / 21/15	650/600 / 24/23	800/750 / 21/15	1750/1500 / 30/7	2250/2250 / 10/10
Grab tensile test — Maximum load (N/25 mm)						
Plane strain tensile test — Tensile strength (kN/m) / Extension at max. load (%)	TRRL SR703	18.3/9.4 / 32/21	16.2/12.5 / 26/29	18.7/18.1 / 33/22.5	36/43 / 26/16	47/45 / 13/13
Puncture and tear						
CBR plunger test — Maximum load (N)		1510	1750	2100	3500	4790
Drop cone test — Hole width (mm) at 0.5 kg/m		8	10	11	4	6
Tear strength — Maximum load (N)	Trapezoidal	370	370	360	855	725
Permeability						
Normal pressure = 0 kPa — Flow rate (l/s/m²) / k(m/s)	(10 cm head)	22	20	20	20	30
Normal pressure = kPa — Flow rate (l/s/m²) / k(m/s)	$\times 10^{-4}$	0.8	0.8	0.7	1.0	2.2
Width(s) available × length (m)		4.5, 5.3 × 100	4.5, 5.3 × 100	4.5, 5.3 × 100	4.5, 5.0 × 100	5.0 × 100

Note: Where two values are given, they refer to the lengthwise/crosswise directions.

		Test reference	Product reference: LOTRAK (HIGH PERMEABILITY)	
			MT22/16	MT22/20
Physical characteristics	Mass per unit area (g/m^2)		120	140
	Thickness (mm)		0.5	0.5
	Pore size (μm) O_{50} O_{90}		740 / 790	470 / 560
Strength	50 mm strip tensile test — Max. load (N/50 mm), Extension at max. load (%)	BS 4859	1100/800 / 25/25	1100/1000 / 25/25
	Grab tensile test — Maximum load (N/25 mm)			
	Plane strain tensile test — Tensile strength (kN/m), Extension at max. load (%)	TRRL SR703	23/18 / 25/25	23/22 / 25/25
Puncture and tear	CBR plunger test — Maximum load (N)		1950	2350
	Drop cone test — Hole width (mm) at 0.5 kg/m		6	5
	Tear strength — Maximum load (N)			
Permeability	Normal to the plane of the geotextile — Normal pressure = kPa: Flow rate (l/s/m^2), k(m/s); Normal pressure = kPa: Flow rate (l/s/m^2), k(m/s)	(10 cm head)	>300 / >10^{-3}	>100 / 10^{-3}
	Width(s) available × length (m)		4.5 × 100	4.5 × 100

Note: Where two values are given, they refer to the lengthwise/crosswise directions.

	Test reference	Product reference: LOTRAK (HEAVY WEIGHTS)					
		50/50	85/85	120/120	200/40	320/60
Physical characteristics	Mass per unit area (g/m^2)		325	500	800	750	1200
	Thickness (mm)		1·4	1·6	2·0	2·0	3·5
	Pore size (μm) O_{90}		240	365	120	145	355
Strength	50 mm strip tensile test	Maximum load (N/50 mm) Extension at max. load (%)	BS 4859	2400/2800 29/15	4700/4200 13/11	6000/6000 13/8	10300/1700 11/7	15600/2900 12/13
	Grab tensile test	Maximum load (N/25 mm)						
	Plane strain tensile test	Tensile strength (kN/m) Extension at max. load (%)	CFG	50/66 16/7	90/87 8/7			318/63 12/12
Puncture and tear	CBR plunger test	Maximum load (N)		5100	14000		8500	
	Drop cone test	Hole width (mm) at 0·5 kgm	(10 cm head)			3		
	Tear strength	Maximum load (N)						
Permeability	Normal to the plane of the geotextile	Normal pressure = kPa Flow rate (l/s/m^2) k(m/s) Normal pressure = kPa Flow rate (l/s/m^2) k(m/s)		27 3·8×10^{-4}	45 7·2×10^{-4}	15 3·0×10^{-4}	12·5 2·5×10^{-4}	20 7·0×10^{-4}
	Width(s) available × length (m)			5×200	5×200	5×100	5×100	5×100

Note: Where two values are given, they refer to the lengthwise/crosswise directions.

	Test reference	Product reference: LOTRAK (COMPOSITES)						
		200	400	400C	600	45/45/200	45/45/600
Physical characteristics	Mass per unit area (g/m²)	320	520	520	720	440	840
	Thickness (mm)	2·5	4·3	2·7	4·6	2·7	4·9
	Pore size (μm)	<30	<30	<30	<30	<30	<30
Strength	50 mm strip tensile test — Max. load (N/50 mm)	580/670	490/670	500/800	470/670	1650/1590	1730/1610
	Extension at max. load (%)	16/13	13/12	11/10	12/12	15/10	22/10
	Grab tensile test — Maximum load (N/25 mm)						
	Plane strain tensile test — Tensile strength (kN/m)						
	Extension at max. load (%)						
Puncture and tear	CBR plunger test — Maximum load (N)	1740	>5000	>5000	>5000	>5000	>5000
	Drop cone test — Hole width (mm) at 1 kgm	9	9	5	0	8	0
	Tear strength — Maximum load (N)						
Permeability	Normal to the plane — Flow rate (l/s/m²) (10 cm head) ×10⁻³	22	20	15	12	10	8
	k(m/s) — Normal pressure = kPa	0·6	0·9	0·4	0·6	0·3	0·4
	Normal pressure of the geotextile = kPa — Flow rate (l/s/m²) k(m/s)						
	Width(s) available × length (m)	4·5×50	4·5×50	4·5×50	4·5×50	5·0×50	5·0×50

Note: Where two values are given, they refer to the lengthwise/crosswise directions.

	Test reference	Product reference: TRAMMEL FIN DRAIN				
		F30	F40	F50	Core	Cut off membrane
Physical characteristics	Mass per unit area (g/m^2)	90	105	120	690	130
	Thickness (mm)	0·4	0·5	0·5	4·0	0·3
	Pore size (μm) O_{50} O_{90}	780 980	560 670	330 420	Mesh 6000×4000	0
Strength	50 mm strip tensile test BS 4859 Max. load (N/50 mm) Extension at max. load (%)	400/1500 20/10	500/1700 20/10	600/1800 30/25		900/700 20/20
	Grab tensile test Maximum load (N/25 mm)					
	Plane strain tensile test Tensile strength (kN/m) Extension at max. load (%)				Yield 4·5/4·5 90/60	Too extensible
Puncture and tear	CBR plunger test Maximum load (N)	1500	1825	1925	Too extensible	
	Drop cone test Hole width (mm)					
	Tear strength Maximum load (N)					
Permeability	Normal pressure = kPa Flow rate (l/s/m^2) k(m/s)	400 1·6	390 1·7	390 1·9	—	0
	Normal pressure = kPa Flow rate (l/s/m^2) k(m/s)				—	0
	Width(s) available × length (m)	3·1×100	3·1×100	3·1×100	2·0×50	7·2×50

Note: Where two values are given, they refer to the lengthwise/crosswise directions.

Compendium of product data

Manufacturer
Du Pont de Nemours International SA
Textile Fibres
Typar Geotextile Group
50–52 route des Acacias
CH-1211 Geneva 24
Switzerland
Telephone: 010 41 22 378111   Telex: 22512 Geneva

UK supplier
Du Pont (UK) Limited
Textile Fibres Department
94 Regent Road
Leicester
LE1 7DJ
Telephone: 0533 470444   Telex: 341191

Product range
Typar 3207, 3337, 3407, 3407–2, 3607–3, 3707, 3857
Spun-bonded continous filament
100% polypropylene

Specify Du Pont TYPAR*
The field-proven geotextile

Whether for civil engineering, road construction, drainage or erosion control applications, Du Pont TYPAR has just the right combination of physical, mechanical and hydraulic properties to do a perfect job.

TYPAR has proven experience, with over 250 million square metres already in use all over the world.

For further information and free samples, please contact:
Du Pont de Nemours International S.A., TYPAR Geotextiles Group, P.O. Box, CH-1211 Geneva, Switzerland
Tel. (022) 37 81 11, Telex 422 512

Control the earth with TYPAR

* Du Pont's registered trademark for its spunbonded polypropylene
Du Pont is a member of the International Geotextile Society

	Test reference	Product reference: TYPAR*						
		3207	3337	3407	3407-2	3607-3	3707	3857
Physical characteristics								
Mass per unit area (g/m^2)		68	110	136	150	190	240	290
Thickness (mm) 2 kPa		0.36	0.45	0.46	0.48	0.56	0.68	0.78
20 kPa		0.33	0.40	0.43	0.44	0.52	0.65	0.75
200 kPa		0.29	0.35	0.39	0.40	0.48	0.63	0.72
Pore size (μm) O_{50}	De Voorst	200	145	125	100	85	60	40
O_{90}		300	190	155	130	110	90	70
O_{95}		340	210	165	140	125	100	85
Strength								
50 mm strip tensile test Max. load (N/50 mm)	DIN 53857	155	250	380	440	515	710	870
Extension at max. load (%)		35	40	40	40	40	40	40
Grab tensile test Maximum load (N/25 mm)	DIN 53858 ASTM D 1682 (Modified)	270	440	565	710	890	1070	1300
		285	470	680	780	1000	1250	1440
Plane strain tensile test Tensile strength (kN/m)	EMPA	3.2	5.2	7.6	9.5	12.0	16.0	19.2
Extension at max. load (%)		30	30	31	38	38	40	40
Puncture and tear								
CBR plunger test Maximum load (N) ($\bar{x}$)	DIN 54307	500	830	1270	1500	1850	2450	3030
Drop cone test Hole width (mm)	SØRLIE	50	36	29	30	23	20	17
Tear strength Maximum load (N)	ASTM D 1117	125	270	370	380	460	570	680
Permeability								
Normal to the plane of the geotextile Normal pressure = kPa Flow rate (l/s/m^2)	De Voorst (10 cm head)	260	160	100	75	60	40	30
k(m/s)								
Normal pressure = kPa Flow rate (l/s/m^2)								
k(m/s)								
Width(s) available × length (m)		2.0×1200	4.25/5.2 ×150	3.5/4.25/5.2 ×150	4.25/5.2 ×150	4.25/5.2 ×100	4.25/5.2 ×100	5.2×100

Note: Where two values are given, they refer to the lengthwise/crosswise directions. * Du Pont's registered trademark for its spunbonded polypropylene.

Manufacturer
Enka Industrial Systems BV
Postbus 306
Velperweg 76
6800 AB Arnhem
The Netherlands
Telephone: 010 31 85 66 44 22   Telex: 45204 enka nl
Telefax: 010 31 85 663727

UK supplier
MMG Civil Engineering Systems
Station Road
Docking
Norfolk
PE31 8LY
Telephone: 048 58 501   Telex: 81586
Telefax: 048 58 741

Product range
Colbond PF150, PF250, PF350
Needlepunched staple fibre chemically bonded
100% polyester

Stabilenka 150/45, 200/45, 300/45, 400/50, 400/100, 600/50, 600/100, 800/100, 1000/100
Multifilament warp, multifilament weft
Warp polyester, weft polyamide

Standard Enkamat is a flexible three-dimensional matting produced by welding filaments of carbon-black nylon at their crossover points.

Enkamat A and *Enkazon* are adaptations of the standard product. Enkamat A incorporates a factory-produced filling of bitumen-bound gravel chippings. Enkazon is a pre-grown turf incorporating standard Enkamat.

Enkadrain is a composite comprising standard Enkamat sandwiched between heat bonded non-woven geotextiles. It is used for structural drainage.

		Test reference	Product reference: COLBOND				STABILENKA	
			PF150	PF250	PF350		150/45	200/45
Physical characteristics	Mass per unit area (g/m²)	DIN 53854	150	250	350		365	450
	Thickness (mm) 2 kN/m² / 200 kN/m²	SN 640550	1·78 / 0·68	2·20 / 1·20	3·0 / 1·83		0·5	0·77
	Pore size (μm) O_{90}	SN 640550	90	90	80		200	340
Strength	50 mm strip tensile test — Max. load (N/50 mm) / Extension at max. load (%)	SN 640550 / DIN 53857	250/300 / 27/64	480/450 / 25/50	600/845 / 45/55		7500/2250 / 9/20	10000/2250 / 9/20
	Grab tensile test — Maximum load (N/25 mm)							
	Plane strain tensile test — Tensile strength (kN/m) / Extension at max. load (%)	SN 640550 / DIN 53857	5·0/6·0 / 27/64	9·6/9·1 / 25/50	12·1/16·9 / 45/55		150/45 / 9/20	200/45 / 9/20
Puncture and tear	CBR plunger test — Maximum load (N)	SN 640550 (DIN 54307)	0·76	1·33	2·16		4000	5500
	Drop cone test — Hole width (mm)	SN 640550	47	37	24		25	16·8
	Tear strength — Maximum load (N)	SN 640550 (DIN 53363)	127/172	151/191	289/391			
Permeability	Normal to the plane of the geotextile — 20 kN/m² Flow rate (l/s/m²) k(m/s) / 200 kN/m² Flow rate (l/s/m²) k(m/s)	SN 640550 (10 cm Head)	$1 \cdot 1 \times 10^{-3}$ / $1 \cdot 9 \times 10^{-4}$	$1 \cdot 3 \times 10^{-3}$ / $4 \cdot 1 \times 10^{-4}$	$1 \cdot 5 \times 10^{-3}$ / $4 \cdot 4 \times 10^{-4}$		11	14
	Width(s) available × length (m)		4·50×400	4·5×250	4·5×200		5·00×300	5·00×300

Note: Where two values are given, they refer to the lengthwise/crosswise directions.

		Test reference	Product reference: STABILENKA						
			300/45	400/50	400/100	600/50	600/100	800/100	1000/800
Physical characteristics	Mass per unit area (g/m^2)	DIN 53854	590	845	945	1205	1335	1590	1960
	Thickness (mm)		—	1-2	—	—	—	—	—
	Pore size (μm) O_{90}	SN 640550	235	73	540	428	400	280	60
Strength	50 mm strip tensile test — Max. load (N/50 mm) / Extension at max. load (%)	SN 640550 / DIN 53857	15000/2250 9/20	20000/5000 10/18	20000/2500 10/18	30000/2500 10/18	30000/5000 10/18	10/18	10/18
	Grab tensile test — Maximum load (N/25 mm)								
	Plane strain tensile test — Tensile strength (kN/m) / Extension at max. load (%)	SN 640550 / DIN 53857	300/45 9/20	400/50 10/18	400/100 10/18	600/50 10/18	600/100 10/18	800/100 10/18	1000/100 10/18
Puncture and tear	CBR plunger test — Maximum load (N)		8150	8400	—	4000	—	—	—
	Drop cone test — Hole width (mm)	SN 640550 (10 cm head)	—	9.5	—	—	—	—	—
	Tear strength — Maximum load (N)		18	2.5	67	25	19	13	2
Permeability	Normal to the plane of the geotextile — 20 kN/m^2 Flow rate (l/s/m^2), k (m/s); 200 kN/m^2 Flow rate (l/s/m^2), k (m/s)								
	Width(s) available × length (m)		4.80×200	5.00×200	5.00×150*	5.00×150*	5.00×150*	5.00×150*	5.00×150*

Note: Where two values are given, they refer to the lengthwise/crosswise directions. *Cut to specified length on request. All values are approximate.

Compendium of product data

Manufacturer
Fibertex APS
Svendborgvej 16
Postbox 8029
9220 Aalborg Ost
Denmark
Telephone: 010 45 8 158600   Telex: 69600 FIBER DK

UK supplier
BSP Limited
Unit 35
Claydon Industrial Park
Gipping Road
Great Blakenham
Ipswich
Suffolk
IP6 0JD
Telephone: 0473 830030   Telex: 98115 BSP G

Product range
Fibertex S-170/S-300/S-400
Fibertex G-100/F-2B/F-3S
Needlepunched staple fibre heat bonded
100% polypropylene

Fibertex F-32M/F-4M
Needlepunched stable fibre
100% polypropylene

		Test reference	Product reference: FIBERTEX			
			S-300	G-100	F-2B	F-3S
Physical characteristics	Mass per unit area (g/m^2)	DIN 53854	300	100	140	230
	Thickness (mm)	DIN 53855	1·3	0·6	0·95	1·2
	Pore size (μm) O_{50} O_{90}		40 60	80 110	60 85	40 55
Strength	50 mm strip tensile test — Max. load (N/50 mm) Extension at max. load (%)	DIN 53857	600/800 40/60	150/150 35/45	400/450 50/60	550/600 60/60
	Grab tensile test — Maximum load (N/25 mm)	DIN 53858	700/1000	270/350	480/530	810/850
	Plane strain tensile test — Tensile strength (kN/m) Extension at max. load (%)		12/16 40/60	4/4 35/45	8/9 50/60	11/12 50/55
Puncture and tear	CBR plunger test — Maximum load (N) ($\bar{s}-s$)	DIN 54307	2200	700	1200	1900
	Drop cone test — Hole width (mm)		20	50	32	19
	Tear strength — Maximum load (N)	ASTM D 1117	300/220	100/70	160/160	270/270
Permeability	Normal to the plane of the geotextile — Normal pressure = 2 kPa — Flow rate (l/s/m^2) k(m/s)	(10 cm head)	19 $2·5 \times 10^{-4}$	150 9×10^{-4}	90 $8·5 \times 10^{-4}$	80 $9·5 \times 10^{-4}$
	Normal pressure = kPa — Flow rate (l/s/m^2) k(m/s)					
	Width(s) available × length (m)		$4·2/5·2 \times 100$	$4·5/5·0 \times 100$	$4·5/5·0 \times 100$	$4·5/5·0 \times 100$

Note: Where two values are given, they refer to the lengthwise/crosswise directions.

	Test reference	Product reference: FIBERTEX	
		F-32M	F-4M
Physical characteristics			
Mass per unit area (g/m²)	DIN 53854	190	300
Thickness (mm)	DIN 53855	2.5	3.2
Pore size (µm) O_{50} O_{90}		70 / 100	50 / 70
Strength			
50 mm strip tensile test — Max. load (N/50 mm) Extension at max. load (%)	DIN 53857	450/500 / 70/70	800/850 / 70/80
Grab tensile test — Maximum load (N/25 mm)	DIN 53858	600/700	1000/1050
Plane strain tensile test — Tensile strength (kN/m) Extension at max. load (%)		9/10 / 70/70	16/17 / 70/70
Puncture and tear			
CBR plunger test — Maximum load (N) ($\bar{x}$ − s)	DIN 54307	1600	2800
Drop cone test — Hole width (mm)		18	13
Tear strength — Maximum load (N)	ASTM D 1117 (10 cm head)	270/270	340/340
Permeability			
Normal to the plane of the geotextile — Normal pressure = 2 kPa: Flow rate (l/s/m²), k(m/s); Normal pressure = kPa: Flow rate (l/s/m²), k(m/s)		150 / 3.7×10^{-3}	105 / 3.5×10^{-3}
Width(s) available × length (m)		$4.5/5.0 \times 100$	$4.5/5.0 \times 100$

Note: Where two values are given, they refer to the lengthwise/crosswise directions.

Manufacturer
Godfreys of Dundee Limited
P.O. Box 46
Dundee
Scotland
DD1 9HN
Telephone: 0382 28866   Telex: 76435

UK supplier
Direct from manufacturer

Product range
Autoway 90, 120, 160
Woven slit film tapes
100% polypropylene

	Test reference		Product reference: AUTOWAY		
			90	120	160
Physical characteristics	Mass per unit area (g/m²)		90	120	160
	Thickness (mm)		0.4	0.45	0.6
	Pore size (μm)	O_{90}	380	230	360
Strength	50 mm strip tensile test	Max. load (N/50 mm) Extension at max. load (%)			
	Grab tensile test	Maximum load (N/25 mm)	613/305	748/812	786/1112
	Plane strain tensile test	Tensile strength (kN/m) Extension at max. load (%)	19/8 22/14	19/15 22/15	19/40 22/10
Puncture and tear	CBR plunger test Burst strength	Maximum load (N) N/C2	165	218	281
	Drop cone test	Hole width (mm)			
	Tear strength	Maximum load (N) 100 mm head	34	14	32
Permeability	Normal to the plane of the geotextile	Normal pressure = kPa Flow rate (l/s/m²) k(m/s) Normal pressure = kPa Flow rate (l/s/m²) k(m/s)			
	Width(s) available × length (m)		4.5×100	4.5×100	4.5×100

Note: Where two values are given, they refer to the lengthwise/crosswise directions.

Manufacturer
Hoechst Aktiengesellschaft
Postfach 800320
D-6230 Frankfurt am Main 80
West Germany
Telephone: 010 49 69 3051   Telex: 003 412340 ho d

UK supplier
Hoechst (UK) Limited
Fibres Division
Holywell Green
Halifax
HX4 9DL
Telephone: 0422 75522   Telex: 51619 huk sta g
Telefax: 0422 71689

Product range
Trevira Spunbond 13/130, 13/150, 11/180, 11/200, 11/270, 11/300, 11/360, 11/500
Needlepunched continuous filament
100% polyester

		Test reference	Product reference: TREVIRA SPUNBOND					
			13/150	11/200	11/270	11/300	11/360	11/500*
Physical characteristics	Mass per unit area (g/m²)	DIN 53854	150	200	270	300	360	500
	Thickness (mm)	DIN 53855	1.7	2.4	2.9	3.1	3.5	4.3
	Pore size (µm) D_w	Franzius-Institut, Hannover	90	160	140	120	110	90
Strength	50 mm strip tensile test — Max. load (N/50 mm) / Extension at max. load (%)	DIN 53857	450/450 / 75/80	560/560 / 65/70	850/850 / 65/70	950/950 / 70/75	1110/1110 / 70/75	1560/1560 / 70/75
	Grab tensile test — Maximum load (N/25 mm)							
	Plane strain tensile test — Tensile strength (kN/m) / Extension at max. load (%)							
Puncture and tear	CBR plunger test — Maximum load (N)	DIN 54307	1330	2002	2900	3198	3689	5671
	Drop cone test — Hole width (mm)							
	Tear strength — Maximum load (N)							
Permeability	Normal to the plane of the geotextile — Normal pressure = kPa, Flow rate (l/s/m²) k(m/s); Normal pressure = kPa, Flow rate (l/s/m²) k(m/s)		28.8×10^{-4} / 1.9×10^{-4}	65.8×10^{-4} / 5.4×10^{-4}	66.0×10^{-4} / 5.8×10^{-4}	69.6×10^{-4} / 6.1×10^{-4}	58.1×10^{-4} / 6.4×10^{-4}	57.3×10^{-4} / 7.7×10^{-4}
	Width(s) available × length (m)			2.20×400 / 5.30×200			2.20×300 / 5.30×200	2.20×250 / 5.30×150

Note: Where two values are given, they refer to the lengthwise/crosswise directions.   *Product data on types 13/130 and 11/180 not available.

Geotextiles handbook

Manufacturer
Huesker Synthetic GmbH & Co
Postfach 1262
D-4423 Gescher
West Germany
Telephone: 010 49 2542 7010   Telex: 89 23 28
Telefax: 010 49 2542 70137

UK supplier
Geotextile Projects Limited
Spirella Building
Bridge Road
Letchworth
Hertfordshire
SG6 4ET
Telephone: 0462 678 448

Product range
HaTe C 00.520, C 50.535, D 00.530
Woven monofilament-on-monofilament
100% polyethylene

HaTe 23.142
Woven multifilament-on-multifilament
100% polyester, PVC coated

HaTe D 00.006
Woven monofilament-on-tape
100% polyethylene

HaTe 43.144, 50.145, 50.145/B
Woven multifilament-on-multifilament
100% polyester, PVC coated

HaTe C 50.002
Woven monofilament-on-multifilament
Warp–polyethylene, weft–polypropylene

HaTe C 10.340
Woven multifilament-on-multifilament
100% polyester

HaTe 60.006, 40.705
Woven fibrillated tape-on-tape
100% polypropylene

HaTelit 20/9, 30/13
Grid: multifilament-on-multifilament
100% polyester, bituminous coating

HaTe 35/20–20, 50/30–20, 80/30–20
Grid: multifilament-on-multifilament
100% polyester, PVC coated

		Test reference	Product reference: HATE							
			C 00.520	C 50.535	D 00.530	23.142	D 00.006	43.144	50.145
Physical characteristics	Mass per unit area (g/m²)		100	220	220	110	130	200	225
	Thickness (mm)								
	Pore size (μm)	DIN 53857	750	350	200	5000	300–450	1200–1500	1000–1200
Strength	50 mm strip tensile test	Max. load (N/50 mm) Extension at max. load (%)		1000/1000 30/20	2200/2100 30/25	3000/1500 30/25	700/700 15/15	1200/1200 25/20	1750/1750 15/15	1900/1900 15/15
	Grab tensile test	Maximum load (N/25 mm)							
	Plane strain tensile test	Tensile strength (kN/m) Extension at max. load (%)							
Puncture and tear	CBR plunger test	Maximum load (N)							
	Drop cone test	Hole width (mm)							
	Tear strength	Maximum load (N)							
Permeability	Normal to the plane of the geotextile	Normal pressure = kPa Flow rate (l/s/m²) k(m/s) Normal pressure = kPa Flow rate (l/s/m²) k(m/s)	10 cm head	500	450	350	>500	190		400
	Width(s) available × length (m)								

Note: Where two values are given, they refer to the lengthwise/crosswise directions.

	Test reference	Product reference: HATE				
		50.145/B	C 50.002	C 10.340	60.006	40.705
Physical characteristics	Mass per unit area (g/m²)	225	225	240	500	750
	Thickness (mm)					
	Pore size (μm)	600–800	60–150	150	200	100
Strength	50 mm strip tensile test — Max. load (N/50 mm) Extension at max. load (%) — DIN 53857	1900/1750 15/15	2200/2500 25/25	3500/3200 20/20	4000/4000 15/15	8500/1500 14/17
	Grab tensile test — Maximum load (N/25 mm)					
	Plane strain tensile test — Tensile strength (kN/m) Extension at max. load (%)					
Puncture and tear	CBR plunger test — Maximum load (N)					
	Drop cone test — Hole width (mm)					
	Tear strength — Maximum load (N)					
Permeability	Normal to the plane of the geotextile — Normal pressure = kPa, Flow rate (l/s/m²), k(m/s); Normal pressure = kPa, Flow rate (l/s/m²), k(m/s) — 10 cm head	100	100	50	15	40
	Width(s) available × length (m)					

Note: Where two values are given, they refer to the lengthwise/crosswise directions.

	Test reference	Product reference: HaTelit		HaTe – GRID		
		20/9	30/13	35/20-20	50/30-20	80/30-20
Physical characteristics	Mass per unit area (g/m²)	260	260	320	450	550
	Thickness (mm)					
	Pore size (μm)			20×20	20×20	20×20
Strength	50 mm strip tensile test — Max. load (N/50 mm) Extension at max. load (%)	20000 × 20000	30000 × 30000	20000 × 20000	20000 × 20000	20000 × 20000
	Grab tensile test — Maximum load (N/25 mm)					
	Plane strain tensile test — Tensile strength (kN/m) Extension at max. load (%) — DIN 53857	50/50	50/50	35/20	50/30	80/30
Puncture and tear	CBR plunger test — Maximum load (N)					
	Drop cone test — Hole width (mm)					
	Tear strength — Maximum load (N)					
Permeability	Normal to the plane of the geotextile — Normal pressure = kPa, Flow rate (l/s/m²), k(m/s); Normal pressure = kPa, Flow rate (l/s/m²), k(m/s)					
	Width(s) available × length (m)	1-7/3-6×150	1-7/3-6×150			

Note: Where two values are given, they refer to the lengthwise/crosswise directions.

Compendium of product data

Manufacturer
ICI Fibres Geotextile Group
Pontypool
Gwent
NP4 0YD
Telephone: 04955 57722   Telex: 498208 ICIFPP G

UK suppliers
Greenham Tool Co. Ltd
Sheffield Insulations Ltd
Macnaughton Blair Ltd
ICI (Ireland) Ltd

Depots throughout the UK and Ireland. Contact manufacturer for further information

Product range
Terram non-wovens
500, 700, 1000, 1500, 2000, 3000
Heat bonded continuous filament
67% polypropylene, 33% polyethylene

NP3, NP4, NP5
Mechanically bonded continuous filament
100% polyester

Terram wovens
W/3-3, W/5-5, W/7-7, W/12-12, W/12-4, W/20-4, W/30-4
Multifilament warp and weft
100% polypropylene

WB/20-5, WB/40-5, WB/40-10, WB/60-5, WB/60-10
Multifilament warp and weft
Polyester warp, polyamide weft

35C
Carbonised monofilament warp and weft
55% polypropylene, 45% polyethylene

Geotextiles handbook

Terram meshes
42 A
100% polyester

Filtram
1B1, 1BZ
Laminated filter drains
Polypropylene/polyethylene

Paraproducts
ParaGrid 50/25S, 50/50S, 100/25S, 100/100S
Polyester/polyethylene

ParaLink 300S up to 1250S
Polyester/polyethylene

ParaWeb 100D
Polyester/polyethylene

	Test reference	Product reference: TERRAM					
		500	700	1000	1500	2000	3000
Physical characteristics							
Mass per unit area (g/m^2)	BS 2471	70	100	135	190	230	270
Thickness (mm)	ICI PTL D 710	0·4	0·5	0·7	0·8	1·0	1·2
Pore size (μm) O_{50} / O_{90}	ICI PTL D 706	200 / 350	120 / 180	70 / 100	40 / 60	30 / 50	30 / 40
Strength							
200 mm strip tensile test — Max. load (kN) / Extension at max. load (%)	ASTM D 4595, BS 6906: Pt 1	0·7 / 20	1·0 / 20	1·6 / 30	2·4 / 30	3·0 / 30	3·6 / 30
Grab tensile test — Maximum load (N/25 mm)	ICI STM 65	330	500	700	1100	1400	1600
Plane strain tensile test — Tensile strength (kN/m) / Extension at max. load (%)							
Puncture and tear							
CBR plunger test — Maximum load (N)	DIN 54307-A	700	1000	1500	2100	2600	3100
Drop cone test — Hole width (mm)							
Tear strength — Maximum load (N)	ASTM D 4533	150	250	300	450	600	700
Permeability							
Normal to the plane of the geotextile — Normal pressure = kPa, Flow rate (l/s/m^2) / k(m/s); Normal pressure = kPa, Flow rate (l/s/m^2) / k(m/s)	ICI PTL D 527A (10 cm head)	150 / 6×10^{-4}	80 / 4×10^{-4}	50 / 3·5×10^{-4}	35 / 2·8×10^{-4}	33 / 3·3×10^{-4}	30 / 3·6×10^{-4}
Width(s) available × length (m)		4·5×200	4·5×150	4·5×100	4·5×100	4·5×100	4·5×100

Note: Where two values are given, they refer to the lengthwise/crosswise directions.

Test reference	Product reference: TERRAM		
	NP3	NP4	NP5

Physical characteristics

	Test reference	NP3	NP4	NP5
Mass per unit area (g/m^2)	BS 2471	200	290	350
Thickness (mm)	ICI PTL D 710	1·5	2·0	2·5
Pore size (μm) O_{50} / O_{90}	ICI PTL D 706	50 / 50	30 / 30	30 / 30

Strength

	Test reference	NP3	NP4	NP5
200 mm strip tensile test — Max load (kN) / Extension at max. load (%)	ASTM D 4595, BS 6906: Pt 1	3·2 / 45	4·4 / 45	5·5 / 45
Grab tensile test — Maximum load (N/25 mm)	ASTM D 1682	800	1000	1450
Plane strain tensile test — Tensile strength (kN/m), Extension at max. load (%)				

Puncture and tear

	Test reference	NP3	NP4	NP5
CBR plunger test — Maximum load (N)	DIN 54307-A	2100	2800	3700
Drop cone test — Hole width (mm)				
Tear strength — Maximum load (N)	ASTM D 4533	850	950	1100

Permeability

	Test reference	NP3	NP4	NP5
Normal pressure = kPa, Flow rate (l/s/m^2), k(m/s); Normal pressure = kPa, Flow rate (l/s/m^2), k(m/s)	ICI PTL D 527A (10 cm head)	160, $2·4 \times 10^{-3}$	90, $1·8 \times 10^{-3}$	60, $1·5 \times 10^{-3}$
Width(s) available × length (m)		5·2 × 125	5·2 × 100	5·2 × 70

Note: Where two values are given, they refer to the lengthwise/crosswise directions.

	Test reference	Product reference: TERRAM						
		W/3-3	W/5-5	W/7-7	W/12-12	W/12-4	W/20-4	W/30-4
Physical characteristics								
Mass per unit area (g/m^2)	BS 2471	170	250	400	550	420	570	720
Thickness (mm)	ICI PTL D 710		1·0	1·3	2·0	1·5	1·8	2·5
Pore size (µm) O_{50} / O_{90}	ICI PTL D 706	250 / 350	250 / 400	250 / 400	250 / 350	200 / 300	200 / 300	200 / 300
Strength								
200 mm strip tensile test — Max load (kN) / Extension at max. load (%)	ASTM D 4595 / BS 6906: Pt 1	6 / 10/5	10 / 9/8	14 / 9/8	24 / 12/8	24/8 / 12/8	40/8 / 12/8	60/8 / 12/8
Grab tensile test — Maximum load (N/25 mm)	ASTM D 1682	—	—	—	—	—	—	—
Plane strain tensile test — Tensile strength (kN/m) / Extension at max. load (%)								
Puncture and tear								
CBR plunger test — Maximum load (N)	DIN 54307-A	2500	3500	4500	8000	—	—	—
Drop cone test — Hole width (mm)								
Tear strength — Maximum load (N)	ASTM D 4533	—	—	—	—	—	—	—
Permeability								
Normal to the plane of the geotextile — Normal pressure = kPa Flow rate (l/s/m^2) k(m/s) / Normal pressure = kPa Flow rate (l/s/m^2) k(m/s)	ICI PTL D 527A (10cm head)	30	30	30	30	30	30	30
Width(s) available × length (m)		4·5×100	4·5×100	4·5×100	4·5×100	4·5×100	4·5×100	4·5×50

Note: Where two values are given, they refer to the lengthwise/crosswise directions.

		Test reference	Product reference: TERRAM					
			WB/20-5	WB/40-5	WB/40-10	WB/60-5	WB/60-10	
Physical characteristics	Mass per unit area (g/m^2)	BS 2471	450	850	950	1200	1350	
	Thickness (mm)	ICI PTL D 710	1·0	1·5	1·7	1·7	2·0	
	Pore size (μm) O_{90}	ICI PTL D 706	30	30	30	30	30	
Strength	200 mm strip tensile test: Max load (kN) / Extension at max. load (%)	ASTM D 4595 / BS 6906: Pt 1	40/10 / 9/15	80/10 / 10/15	80/20 / 10/18	120/10 / 10/18	120/20 / 10/18	
	Grab tensile test: Maximum load (N/25 mm)							
	Plane strain tensile test: Tensile strength (kN/m) / Extension at max. load (%)							
Puncture and tear	CBR plunger test: Maximum load (N)							
	Drop cone test: Hole width (mm)							
	Tear strength: Maximum load (N)							
Permeability	Normal to the plane of the geotextile: Normal pressure = kPa, Flow rate (l/s/m^2), k(m/s); Normal pressure = kPa, Flow rate (l/s/m^2), k(m/s)	ICI PTL D 527A (10 cm head)	35	30	25	30	25	
	Width(s) available × length (m)		5·2×100	5·2×100	5·2×100	5·2×100	5·2×100	

Note: Where two values are given, they refer to the lengthwise/crosswise directions.

		Test reference	Product reference: TERRAM 35C
Physical characteristics	Mass per unit area (g/m^2)	BS 2471	230
	Thickness (mm)		
	Pore size (μm) O_{50} O_{90}	ICI PTL D 706	180 200
Strength	200 mm strip tensile test — Max. load (kN) / Extension at max load (%)	ASTM D 4595 BS 6906: Pt 1	9 25
	Grab tensile test — Maximum load (N/25 mm)		
	Plane strain tensile test — Tensile strength (kN/m) / Extension at max. load (%)		
Puncture and tear	CBR plunger test — Maximum load (N)	DIN 54307-A	6000
	Drop cone test — Hole width (mm)		
	Tear strength — Maximum load (N)		
Permeability	Normal to the plane of the geotextile — Normal pressure = kPa — Flow rate (l/s/m^2) k(m/s); Normal pressure = kPa — Flow rate (l/s/m^2) k(m/s)	ICI PTL D 527A (10 cm head)	90
	Width(s) available × length (m)		3·5 × 100

Note: Where two values are given, they refer to the lengthwise/crosswise directions.

Product reference: TERRAM / FILTRAM

		Test reference	TERRAM 42A	FILTRAM 1B1	FILTRAM 1BZ
Physical characteristics	Mass per unit area (g/m²)	BS 2471	100	1000	1000
	Thickness (mm)				
	Pore size (μm) O_{50} / O_{90}	ICI PTL D 706	4000 / 4000	70 / 100	70 / 100
Strength	200 mm strip tensile test — Max. load (kN) / Extension at max load (%)	ASTM D 4595 / BS 6906: Pt 1	2·8 / 15	3·4 / 20	3·4 / 20
	Grab tensile test — Maximum load (N/25 mm)				
	Plane strain tensile test — Tensile strength (kN/m) / Extension at max. load (%)				
Puncture and tear	CBR plunger test — Maximum load (N)				
	Drop cone test — Hole width (mm)				
	Tear strength — Maximum load (N)	ICI PTL D527A (10 cm head)	500		
Permeability	Normal to the plane of the geotextile — Normal pressure = kPa / Flow rate (l/s/m²) / k(m/s) / Normal pressure = kPa / Flow rate (l/s/m²) / k(m/s)			36	36/0
	Width(s) available × length (m)		3·8 × 100	1·9 × 25	1·9 × 25

Note: Where two values are given, they refer to the lengthwise/crosswise directions.

	Test reference	Product reference:							
		PARAGRID				PARALINK			PARAWEB
		50/25S	50/50S	100/25S	100/100S	300S	1250S	100D
Physical characteristics									
Mass per unit area (g/m²)	BS 2471	400	530	520	770	1660	4350	1750
Thickness (mm)	ICI PTL D 710	2·5	2·5	2·7	3·0	5	10	5
Pore size (μm)								
Strength									
200 mm strip tensile test — Max. load (N/50 mm) / Extension at max. load (%)	ASTM D 4595 / BS 6906: Pt 1	10/5 / 10	10/10 / 10	20/5 / 10	20/20 / 10	60 / 9	250 / 9	10 / –
Grab tensile test — Maximum load (N/25 mm)								
Plane strain tensile test — Tensile strength (kN/m) / Extension at max. load (%)								
Puncture and tear									
CBR plunger test — Maximum load (N)								
Drop cone test — Hole width (mm)								
Tear strength — Maximum load (N)								
Permeability									
Normal to the plane of the geotextile — Normal pressure = kPa, Flow rate (l/s/m²), k(m/s); Normal pressure = kPa, Flow rate (l/s/m²), k(m/s)								
Width(s) available × length (m)		4·5×50	4·5×50	4·5×50	4·5×50	4·5×100	4·5×100	4·4×30

Note: Where two values are given, they refer to the lengthwise/crosswise directions.

Geotextiles handbook

Manufacturer
LEC Limited
Clarendon House
North Station Road
Colchester
Essex
CO1 1UX
Telephone: 0206 42108   Telex: 987908 LECL G
Telefax: 0206 42130

UK supplier
Direct from manufacturer

Product range
Terra-Lec 215WF, 250WF
Woven monofilament
100% polyethylene

Terra-Lec 230WF, 300WF
Woven polyethylene monofilament on polyester multifilament

Terra-Lec 550 WF
Woven multifilament
100% polyester

Terra-Lec 135WT, 175WT, 350WT, 435WT, 500WT
Woven tape
100% polypropylene

Terra-Lec 200NW, 280NW, 350NW, 400NW, 600NW
Needlepunched
100% polypropylene

Geonet GN1 and GN2
Data on application

		Test reference	Product reference: TERRA-LEC WF					
			215WF	230WF	250WF	300WF	550WF
Physical characteristics	Mass per unit area (g/m^2)	DIN 53854	215	230	250	300	550
	Thickness (mm)	DIN 53855	0·82	0·75	0·80	0·65	0·85
	EOS* Pore size (μm)	CW 02215	800	400	310	220	100
Strength	50 mm strip tensile test	Max. load (N/50 mm) Extension at max. load (%)					
	Grab tensile test	Maximum load (N/50 mm)	DIN 53858	2200	2200	3400/1800	3400	9000/4000
	Plane strain tensile test	Tensile strength (kN/m) Extension at max. load (%)					
Puncture and tear	CBR plunger test	Maximum load (N)					
	Drop cone test	Hole width (mm)					
	Tear strength	Maximum load (N)					
Permeability	Normal to the plane of the geotextile	Normal pressure = kPa Flow rate (l/s/m^2) k(m/s)	(10 cm head)	3000	400	400	100	3
		Normal pressure = kPa Flow rate (l/s/m^2) k(m/s)						
	Width(s) available × length (m)		5·05 × 100	5·05 × 100	5·05 × 100	5·05 × 100	5·05 × 100

Note: Where two values are given, they refer to the lengthwise/crosswise directions.   *Equivalent opening size — approximates to 0_{90}.

	Test reference	Product reference: TERRA-LEC WT				
		135WT	175WT	350WT	435WT	500WT
Physical characteristics						
Mass per unit area (g/m^2)	DIN 53854	135	175	350	460	550
Thickness (mm)	DIN 53855	0·40	0·50	1·28	1·00	1·60
EOS* Pore size (µm)	CW 02215	155	310	190	90	150
Strength						
50 mm strip tensile test	Max. load (N/50 mm) Extension at max. load (%)					
Grab tensile test	Maximum load (N/50 mm)	1100	2200	2800	4000/1500	4400
Plane strain tensile test	Tensile strength (kN/m) Extension at max. load (%)					
Puncture and tear						
CBR plunger test	Maximum load (N)					
Drop cone test	Hole width (mm)					
Tear strength	Maximum load (N)					
Permeability						
Normal to the plane of the geotextile	Normal pressure = kPa Flow rate (l/s/m^2) k(m/s) Normal pressure = kPa Flow rate (l/s/m^2) k(m/s)	15 (10 cm head)	17	20	8	12
Width(s) available × length (m)		5·05×100	5·05×100	5·05×100	5·05×100	5·05×100

Note: Where two values are given, they refer to the lengthwise/crosswise directions.

		Test reference	Product reference: TERRA-LEC NW					
			200NW	280NW	350NW	400NW	600NW	
Physical characteristics	Mass per unit area (g/m^2)	DIN 53854	200	280	350	400	600	
	Thickness (mm)	DIN 53855	2·6	3·3	3·9	4·2	6·0	
	EOS* Pore size (μm)	SW 02215	100–150	100–150	100–150	100–150	100–150	
Strength	50 mm strip tensile test	Max. load (N/50 mm) Extension at max. load (%)						
	Grab tensile test	Maximum load (N/25 mm)						
	Plane strain tensile test	Tensile strength (kN/m) Extension at max. load (%)						
Puncture and tear	CBR plunger test	Maximum load (N)						
	Drop cone test	Hole width (mm)						
	Tear strength	Maximum load (N)	ASTM D 2263	180	360	520	630	1100
Permeability	Normal to the plane of the geotextile	Normal pressure = kPa Flow rate (l/s/m^2) k(m/s)						
		Normal pressure = kPa Flow rate (l/s/m^2) k(m/s)	115	91	77	71	50	
	Width(s) available × length (m)		2·9/5·9×100	2·9/5·9×100	2·9/5·9×100	2·9/5·9×100	2·9/5·9×100	

Note: Where two values are given, they refer to the lengthwise/crosswise directions.   * Equivalent opening size — approximates to 0_{90}.

Manufacturer
Malcolm, Ogilvie and Company Limited
Constable Works
31 Constitution Street
Dundee
Scotland
DD3 6NL
Telephone: 0382 22974

UK supplier
Direct from manufacturer

Product range
Wyretex 1-9
Woven multifilament-on-multifilament, reinforced in both warp and weft directions with monofilament steel wire
Either polypropylene/wire or jute/wire

	Test reference	Product reference: WYRETEX					
		1(P/W)*	1(J/W)†	2(P/W)	2(J/W)	3(P/W)	4(P/W)	5(P/W)

			Product reference: WYRETEX					
			1(P/W)*	1(J/W)†	2(P/W)	2(J/W)	3(P/W)	4(P/W)	5(P/W)
Physical characteristics	Mass per unit area (g/m²)		1006	1198	865	1035	709	638	567
	Thickness (mm)								
	Pore size (µm)								
Strength	50 mm strip tensile test	Max. load (N/50 mm) Extension at max. load (%)	3120/2450	2450/2090	2450/2450	2090/2090	1870/2450	1870/2140	1870/1870
	Grab tensile test	Maximum load (N/25 mm)							
	Plane strain tensile test	Tensile strength (kN/m) Extension at max. load (%)							
Puncture and tear	CBR plunger test	Maximum load (N)							
	Drop cone test	Hole width (mm)							
	Tear strength	Maximum load (N)							
Permeability	Normal to the plane of the geotextile	Normal pressure = kPa Flow rate (l/s/m²) k(m/s) Normal pressure = kPa Flow rate (l/s/m²) k(m/s)							
	Width(s) available × length (m)		1·5×50	1·5×50	1·5×50	1·5×50	1·5×50	1·5×50	1·5×50

Note: Where two values are given, they refer to the lengthwise/crosswise directions. * Polypropylene/wire. † Jute/wire.

Test reference		Product reference: WYRETEX					
		6(P/W)	7(P/W)	8(P/W)	8(J/W)	9(P/W)	9(J/W)
Physical characteristics	Mass per unit area (g/m^2)	496	425	284	340	199	234
	Thickness (mm)						
	Pore size (μm)						
Strength	50 mm strip tensile test: Max. load (N/50 mm) Extension at max. load (%)	1870/1650	1650/1650	740/740	720/720	510/510	490/490
	Grab tensile test: Maximum load (N/25 mm)						
	Plane strain tensile test: Tensile strength (kN/m) Extension at max. load (%)						
Puncture and tear	CBR plunger test: Maximum load (N)						
	Drop cone test: Hole width (mm)						
	Tear strength: Maximum load (N)						
Permeability	Normal to the plane of the geotextile: Normal pressure = kPa, Flow rate (l/s/m^2), k(m/s); Normal pressure = kPa, Flow rate (l/s/m^2), k(m/s)						
	Width(s) available × length (m)	1·5×100	1·5×100	1·5×100	1·5×100	1·5×100	1·5×100

Note: Where two values are given, they refer to the lengthwise/crosswise directions. * Polypropylene/wire. † Jute/wire.

Compendium of product data

Manufacturer
Monsanto Europe SA/NV
270–272 Avenue de Tervuren
B-1150 Brussels
Belgium

UK supplier
Monsanto plc
Thames Tower
Burleys Way
Leicester
LE1 3TP
Telephone: 0533 20864

Product range
Hydraway Drain is a high flow, high crush resistant fin drain comprising a non-woven needlepunched geotextile permanently bonded to both sides of an internal supporting core.

Walldrain is similar in concept to *Hydraway Drain*, but with the geotextile bonded to one side only.

Geotextiles handbook

Manufacturer
Netlon Limited
Kelly Street
Blackburn
BB2 4PJ
Telephone: 0254 62431   Telex: 63313   Telefax: 680008

UK supplier
Direct from manufacturer or by way of UK distributors:
Frank Parker & Co. Ltd
In Northern Ireland and Eire: TAL Stabilisation

Product range
Tensar* products
SS1, SS2, SS3, SS35
Biaxial geogrids used for ground stabilisation
100% polypropylene

SR2, SR55, SR80, SR110
Uniaxial geogrid used for soil reinforcement
100% high density polyethylene

AR1
Biaxial geogrids used for asphalt reinforcement
100% polypropylene

GM4
Biaxial geogrid used for gabions and mattresses
100% high density polyethylene

CR1, CR2
Biaxial geogrid used for concrete
100% polypropylene

Mat
Uniaxial geogrid used for erosion control
100% polyethylene

---

*Tensar is a registered trademark of Netlon Limited in the UK and in other countries.

Compendium of product data

Netlon* products
All of these products are integrally extruded meshes.

PW1 — used for pipe wrapping
Cellular medium density polyethylene

CE131 — used for turf reinforcement
100% high density polyethylene

CE151 — a layflat tube used for tubular gabions
100% high density polyethylene

---

*Netlon is a registered trademark of Netlon Limited in the UK and in other countries.

	Test reference	Product reference: TENSAR*						
		SS1	SS2	SS3	SS35	SR2	SR55	SR80
Physical characteristics	Mass per unit area (g/m²)	200	300	240	550	850	500	700
	Thickness (mm)							
	Pore size (μm)	28000×38000	28000×40000	51000×71000	32000×29000	111000×22000	140000×16000	140000×16000
Strength	50 mm strip tensile test Max. load (N/50 mm) Extension at max. load (%)							
	Grab tensile test Maximum load (N/25 mm)							
	Plane strain tensile test Tensile strength (kN/m) Extension at max. load (%)	12·5/20·5 14·0/13·8	17·5/31·5 12·0/11·0	16·1/27·4 16·5/10·6	42·0/34·0 14·0/14·0	80 10·5	55 13·0	80 11·2
Puncture and tear	CBR plunger test Maximum load (N)							
	Drop cone test Hole width (mm)							
	Tear strength Maximum load (N)							
Permeability	Normal pressure = kPa to the plane of the geotextile Flow rate (l/s/m²) k(m/s) Normal pressure = kPa Flow rate (l/s/m²) k(m/s)							
	Width(s) available × length (m)	4×50	4×50	4×50	3·5×50	1×30	1×30	1×30

Note: Where two values are given, they refer to the lengthwise/crosswise directions.

	Test reference	Product reference: TENSAR					
		SR110	AR1	GM4	CR1	CR2	MAT
Physical characteristics	Mass per unit area (g/m^2)	1100	250	320	260	200	450
	Thickness (mm)						18
	Pore size (μm)	132000×16000	51000×71000	62000×62000	35000×35000	65000×65000	—
Strength	50 mm strip tensile test	Max. load (N/50 mm)					
		Extension at max. load (%)					
	Grab tensile test	Maximum load (N/25 mm)					
	Plane strain tensile test	Tensile strength (kN/m)	110	14·0/18·0	15·0/15·0	21·0/21·0	18·0/18·0	3·0
		Extension at max. load (%)	11·2	14·0/10·0	19·0/15·0	12·0/9·0	10·0/10·0	—
Puncture and tear	CBR plunger test	Maximum load (N)					
	Drop cone test	Hole width (mm)					
	Tear strength	Maximum load (N)					
Permeability	Normal to the plane of the geotextile	Normal pressure = kPa	Flow rate (l/s/m^2)					
			k(m/s)					
		Normal pressure = kPa	Flow rate (l/s/m^2)					
			k(m/s)					
	Width(s) available × length (m)	1×30	4×50	4×50	3·5×50	3·5×50	1·5/3·0/4·5×30

Note: Where two values are given, they refer to the lengthwise/crosswise directions.

	Test reference	Product reference: NETLON		
		PW1	CE131	CE151*
Physical characteristics	Mass per unit area (g/m^2)	820	660	1100
	Thickness (mm)			
	Pore size (μm)	9000× 9000	27000× 27000	74000× 74000
Strength	50 mm strip tensile test — Max. load (N/50 mm) / Extension at max. load (%)			
	Grab tensile test — Maximum load (N/25 mm)			
	Plane strain tensile test — Tensile strength (kN/m) / Extension at max. load (%)	— / —	5·80 / 16·5	4·82 / 23·2
Puncture and tear	CBR plunger test — Maximum load (N)			
	Drop cone test — Hole width (mm)			
	Tear strength — Maximum load (N)			
Permeability	Normal to the plane of the geotextile — Normal pressure = kPa / Flow rate (l/s/m^2) / k(m/s) / Normal pressure = kPa / Flow rate (l/s/m^2) / k(m/s)			
	Width(s) available × length (m)	1·63×30	2×30	—

Note: Where two values are given, they refer to the lengthwise/crosswise directions.

*Product is a 1 m wide layflat tube 15 m in length.

Compendium of product data

Manufacturer
Nicolon BV
Postbus 236
7600 AE Almelo
The Netherlands
Telephone: 010 31 5490 44811   Telex: 44440
Telefax: 010 31 5490 44490

UK suppliers
MMG Civil Engineering Systems
Station Road
Docking
Norfolk
PE31 8LY
Telephone: 048 58 501   Telex: 81586
Telefax: 048 58 741

Product range
Geolon 15, 40
Woven slit film tape
100% polypropylene

Geolon 80
Woven multifilament
100% polyester

Geolon 200
Woven fibrillated tape
100% polypropylene

Nicolon H25, H35, H50
Woven fibrillated tape
100% polypropylene

Nicolon M36, M39, M47
Woven monofilament
100% polyethylene

Geotextiles handbook

Nicolon F24, F30
Woven monofilament warp, slit film weft
100% polyethlyene

Nicolon F40
Woven monofilament warp, multifilament weft
Warp polyethylene, weft polypropylene

Test reference		15	40	80	200	H25	H35	H50
		\multicolumn{4}{c}{Product reference: GEOLON}	\multicolumn{3}{c}{NICOLON}

	Test reference	GEOLON 15	GEOLON 40	GEOLON 80	GEOLON 200	NICOLON H25	NICOLON H35	NICOLON H50
Physical characteristics	Mass per unit area (g/m^2)	100	190	360	730	730	325	500
	Thickness (mm)	0.4	0.65	1.30	2.30	2.30	1.5	1.6
	Pore size (μm) O_{90}	150	125	250	250	250	260	213
Strength	50 mm strip tensile test — Max. load (N/50 mm), Extension at max. load (%)							
	Grab tensile test — Maximum load (N/25 mm)							
	Plane strain tensile test — Tensile strength (kN/m)	15/15	40/36	80/50	200/40	200/40	50/55	80/80
	Extension at max. load (%)	20/10	15/16	20/10	15/8	15/8	23/10	15/10
Puncture and tear	CBR plunger test — Maximum load (N)	2000	5000	10500	10370	10370	5100	10700
	Drop cone test — Hole width (mm)	11	8	8	6	6	6	6
	Tear strength — Maximum load (N)	350/300	500/400	1200/600	3500/400	3500/400	800/650	1200/900
Permeability	Normal to the plane of the geotextile — 10 cm head — Flow rate (l/s/m^2) Normal pressure = kPa; k (m/s); Flow rate (l/s/m^2) Normal pressure = kPa; k(m/s)	5	20	31	10	10	30	25
	Width(s) available × length (m)	5.15×200	5.2×200	5.0×200	5.0×200	5.0×200	5.0×200	5.1×200

Note: Where two values are given, they refer to the lengthwise/crosswise directions.

	Test reference	Product reference: NICOLON						
			M36	M39	M47	F24	F30	F40
Physical characteristics	Mass per unit area (g/m²)		115	210	200	235	165	135
	Thickness (mm)		0·55	0·80	0·90	0·70	0·70	0·85
	Pore size (μm)	O_{90}	385	450	1000	180	300	250
Strength	50 mm strip tensile test	Max. load (N/50 mm) Extension at max. load (%)						
	Grab tensile test	Maximum load (N/25 mm)						
	Plane strain tensile test	Tensile strength (kN/m) Extension at max. load (%)	25/25 25/25	40/40 30/20	40/35 30/20	40/40 23/20	40/20 20/20	20/20 23/10
Puncture and tear	CBR plunger test	Maximum load (N)	2890	4030	3930	6190	4990	2500
	Drop cone test	Hole width (mm)	18	12	12	8	13	9
	Tear strength	Maximum load (N)	500/400	900/500	800/600	700/500	800/300	400/400
Permeability	Normal to the plane of the geotextile	Normal pressure = kPa Flow rate (l/s/m²) k(m/s) Normal pressure = kPa Flow rate (l/s/m²) k(m/s)	10 cm head 730	610	630	45	100	35
	Width(s) available × length (m)		5·0×100	5·0×100	5·0×100	5·0×100	5·0×100	5·2×200

Note: Where two values are given, they refer to the lengthwise/crosswise directions.

Compendium of product data

Manufacturer
Rhône-Poulenc Fibres
Department Nontisseé BIDIM
44 rue Salvador Allende
BP 80 – 95871 Bezons Cedex
France
Telephone: 010 33 1 39 47 33 40   Telex: 697802 F

UK supplier
Rhône-Poulenc (UK) Limited
271 High Street
Uxbridge
Middlesex
UB8 1LQ
Telephone: 0895 74080   Telex: 28184

Monomet Limited
50 Beddington Lane
Croydon
CR0 4TE
Telephone: 01 689 8990   Telex: 945486

Product range
Bidim B1, B2, B3, B4, B5, B6, B7
Needlepunched continuous filament
100% polyester

		Test reference	Product reference: BIDIM						
			B1	B2	B3	B4	B5	B6	B7
Physical characteristics	Mass per unit area (g/m^2)	NF G38013 DIN 53854	110	140	160	180	220	260	340
	Thickness (mm) 　2 kPa 　20 kPa 　200 kPa	NF G38012 SN 640550 DIN 53855	1·4 0·8 0·4	1·7 1·0 0·5	1·8 1·1 0·6	1·9 1·2 0·7	2·4 1·4 0·8	2·5 1·6 0·9	2·8 2·1 1·3
	Pore size (μm) 　O_{90} 　D_w	University of Liège Franzius-Institut	235 140	230 140	220 130	220 130	212 130	210 120	150 120
Strength	50 mm strip tensile test	Max. load (N/50 mm) Extension at max. load (%)	NF G07001 DIN 53857/1	230 60–80	340 60–80	420 50–70	475 50–70	600 50–65	720 50–65	980 50–65
	Grab tensile test	Maximum load (N/25 mm)	NF G07120 DIN 53858	510	680	800	900	1100	1320	1700
	Plane strain tensile test	Tensile strength (kN/m) Extension at max. load (%)	NF G38014	7·0 48	9·0 48	12·0 48	13·0 43	17·0 43	21·0 40	26·0 40
Puncture and tear	CBR plunger test	Maximum load (N)	DIN 54307 SN 640550	911 1000	1360 1370	1740 1640	2100 1840	2814 2430	3307 3200	3900 4080
	Drop cone test	Hole width (mm)	SN 640550 Veglaboratoriet OM (10 cm head)	32 31	29 29	28 28	26 26	23 24	20 21	16 16
	Tear strength	Maximum load (N)	DIN 53363/BS4303	160	240	290	350	440	560	760
Permeability	Normal pressure = 20 kPA	Flow rate (l/s/m^2) k(m/s)	NF G38017 SN 640550	190 1·6×10^{-3}	170 1·5×10^{-3}	140 1·3×10^{-3}	130 1·2×10^{-3}	120 1·7×10^{-3}	120 1·9×10^{-3}	110 2·3×10^{-3}
	Normal pressure = 200 kPa	Flow rate (l/s/m^2) k(m/s)		150 0·7×10^{-3}	120 0·6×10^{-3}	100 0·5×10^{-3}	90 0·4×10^{-3}	90 0·5×10^{-3}	80 0·6×10^{-3}	70 0·8×10^{-3}
	Width(s) available × length (m)		All grades 2·1 m, 4·2 m and 5·3 m wide × 150 m long							

Note: Where two values are given, they refer to the lengthwise/crosswise directions.

Geotextiles handbook

Manufacturer
SA UCO NV
Industrial Fabrics Division
Bellevue 1
B-9218 Gent
Belgium
Telephone: 010 32 91 30 90 50   Telex: 11450 UCO B
Telefax: 091/31 73 58

UK supplier
Armco Construction Products Limited
Stephenson Street
Newport
Gwent
NP9 0XH
Telephone: 0633 273081   Telex: 498210

Product range
UCO geotextiles
Standard grades
SG16/16, 22/22L, 22/22M, 25/25, 34/29, 34/34, 42/45, 56/60, 52/52, 74/74, 80/50, 90/90, 200/40
Tape-on-tape
100% polypropylene

SG 22/22H
Tape-on-monofilament
Polypropylene warp, polyethylene weft

SG 70/70, 200/84
Multifilament-on-multifilament
100% polyester

High flow grades
HF 300, 350, 360, 460, 500, 1200, 1300
Monofilament-on-monofilament
100% polyethylene

Compendium of product data

HF 80, 180, 240, 250, 290
Monofilament-on-tape
Polyethylene warp, polypropylene weft

HF 190
Tape-on-monofilament
Polypropylene warp, polyethylene weft

HF 255
Monofilament-on-multifilament
Polyethylene warp, polyester weft

High strength grades
HS 80/60, 80/80, 100/60, 120/40, 120/60, 130/60, 150/45, 150/60, 200/45, 300/45, 400/50, 400/100, 600/50, 600/100, 800/100, 1000/100
Multifilament-on-multifilament
100% polyester

Reflective cracking grades
RC 30/30, 60/60
Multifilament-on-multifilament
100% polyester

Linerguard grades
LG 10/10, 30/30
Composite geotextile comprising woven tape-on-tape fabric with needlepunched fibres
100% polypropylene

Non-woven grades
NW 150, 200, 250, 300, 350, 400, 450
Needlepunched continuous filament
100% polyester

		Test reference	Product reference: UCO GEOTEXTILES (Standard grades)							
			SG 16/16	SG 22/22L	SG 22/22M	SG 22/22H	SG 25/25	SG 34/29	SG 34/34
Physical characteristics	Mass per unit area (g/m^2)		105	140	140	120	130	143	180
	Thickness (mm)		0·6	0·65	0·65	0·65	0·50	0·60	0·80
	Pore size (μm) O$_{90}$		150	130	230	200	300	130	120
Strength	50 mm strip tensile test	Max. load (N/50 mm) Extension at max. load (%)	800/800 20/20	1100/1100 20/20	1100/1100 20/20	1100/1100 20/20	1250/1250 25/25	1700/1450 12/10	1700/1700 15/15
	Grab tensile test	Maximum load (N/25 mm)							
	Plane strain tensile test	Tensile strength (kN/m) Extension at max. load (%)	16/16	22/22	22/22	22/22	25/25	34/29	34/34
Puncture and tear	CBR plunger test	Maximum load (N)							
	Drop cone test	Hole width (mm)							
	Tear strength	Maximum load (N) ASTM D-1117-80	250/350	300/300	300/300	300/500	680/740	510/510	900/900
Permeability	Normal to the plane of the geotextile	Flow rate (l/s/m^2) Normal pressure = kPa k(m/s) Flow rate (l/s/m^2) Normal pressure = kPa k(m/s) (10 cm head)	15	15	28	50	90	10	15
	Width(s) available × length (m)								

Note: Where two values are given, they refer to the lengthwise/crosswise directions.

	Test reference	Product reference: UCO GEOTEXTILES (Standard grades)							
		SG 42/45	SG 52/60	SG 52/52	SG 70/70	SG 74/74	SG 80/50	SG 90/90
Physical characteristics	Mass per unit area (g/m²)		203	270	315	220	525	355	500
	Thickness (mm)		0·70	1·20	1·10	0·40	1·80	1·50	1·80
	Pore size (µm)		130	350	220	120	150	400	260
Strength	50 mm strip tensile test	Max. load (N/50 mm)	2100/2250	2600/3000	2600/2600	3500/3500	3700/3700	4000/2500	4500/4500
		Extension at max. load (%)	12/10	20/8	25/20	11/14	8/8	8/8	8/8
	Grab tensile test	Maximum load (N/25 mm)							
	Plane strain tensile test	Tensile strength (kN/m)	42/45	52/60	52/52	70/70	74/74	80/50	90/90
Puncture and tear	CBR plunger test	Maximum load (N)							
	Drop cone test	Hole width (mm)							
	Tear strength	Maximum load (N) ASTM D-1117-80	535/535	1250/1250	800/800	1100/1100	1200/1000	1400/600	1500/1500
Permeability	Normal to the plane of the geotextile	Normal pressure = kPa	10	35	40	5	10	30	125
		Flow rate (l/s/m²) (10 cm head)							
		k (m/s)							
		Normal pressure = kPa							
		Flow rate (l/s/m²)							
		k (m/s)							
	Width(s) available × length (m)								

Note: Where two values are given, they refer to the lengthwise/crosswise directions.

		Test reference	Product reference: UCO GEOTEXTILES (Standard grades and high flow grades)					
			SG 200/40	SG 200/84	HF 80	HF 180	HF 190	HF 240
Physical characteristics	Mass per unit area (g/m^2)		750	540	220	225	220	225
	Thickness (mm)		2·60	1·80	0·80	0·70	0·80	0·85
	Pore size (μm) O_{90}		200	100	80	180	190	240
Strength	50 mm strip tensile test	Max. load (N/50 mm) Extension at max. load (%)	10000/2000 8/8	10000/4200 20/20	3000/1500 15/15	2700/2000 25/25	2400/1800 25/25	2500/1800 25/15
	Grab tensile test	Maximum load (N/25 mm)						
	Plane strain tensile test	Tensile strength (kN/m) Extension at max. load (%)	200/40	200/84	60/30	54/40	48/36	50/36
Puncture and tear	CBR plunger test	Maximum load (N)						
	Drop cone test	Hole width (mm)						
	Tear strength	Maximum load (N)	3200/550	3200/1150	1200/400	1200/800	650/650	800/650
Permeability	Normal to the plane of the geotextile	Normal pressure = kPa Flow rate (l/s/m^2) k(m/s) Normal pressure = kPa Flow rate (l/s/m^2) k(m/s)	ASTM D-1117-80 (10 cm head)	10	3	15	50	50	50
	Width(s) available × length (m)							

Note: Where two values are given, they refer to the lengthwise/crosswise directions.

	Test reference	Product reference: UCO GEOTEXTILES (High flow grades)								
		HF 250	HF 255	HF 290	HF 300	HF 350	HF 360	HF 460		
Physical characteristics	Mass per unit area (g/m^2)	220	290	275	125	208	210	200		
	Thickness (mm)	0·80	0·70	0·90	0·50	0·80	0·80	0·70		
	Pore size (μm) 0_{90}	250	255	290	300	350	360	460		
Strength	50 mm strip tensile test — Max. load (N/50 mm) / Extension at max. load (%)	2700/1350 25/20	2700/3400 25/15	2700/2000 25/15	1650/1300 25/20	3000/1470 25/20	2700/1700 25/20	2100/2100 25/20		
	Grab tensile test — Maximum load (N/25 mm)									
	Plane strain tensile test — Tensile strength (kN/m) / Extension at max. load (%)	54/27	54/68	54/40	33/26	60/29.4	54/34	42/42		
Puncture and tear	CBR plunger test — Maximum load (N)									
	Drop cone test — Hole width (mm)									
	Tear strength — Maximum load (N) ASTM D-1117-80									
Permeability	Normal to the plane of the geotextile — Normal pressure = kPa, Flow rate (l/s/m^2), k(m/s)	1400/900	1350/1150	1550/650	500/400	1100/600	1200/550	950/650		
	Normal pressure = kPa, Flow rate (l/s/m^2), k(m/s) (10 cm head)	50	100	85	600	550	560	900		
	Width(s) available × length (m)									

Note: Where two values are given, they refer to the lengthwise/crosswise directions.

	Test reference	Produce reference: UCO GEOTEXTILES (High flow and high strength grades)					
		HF 500	HF 1200	HF 1300	HS 80/60	HS 80/80	HS 100/60
Physical characteristics	Mass per unit area (g/m^2)	185	195	90	250	280	285
	Thickness (mm)	0-70	0-85	0-40			
	Pore size (µm) O_{90}	500	1200	1300			
Strength	50 mm strip tensile test — Max. load (N/50 mm)	2100/1900	2100/1850	900/900	4000/3000	4000/4000	5000/3000
	Extension at max. load (%)	25/20	20/15	20/15	10/20	10/20	10/20
	Grab tensile test — Maximum load (N/25 mm)						
	Plane strain tensile test — Tensile strength (kN/m)	42/38	42/37	18/18	80/60	80/80	100/60
	Extension at max. load (%)						
Puncture and tear	CBR plunger test — Maximum load (N)						
	Drop cone test — Hole width (mm)						
	Tear strength — Maximum load (N) ASTM D-1117-80	950/500	900/800	450/350			
Permeability	Normal to the plane of the geotextile — Normal pressure = kPa — Flow rate (l/s/m^2) k(m/s)	1000	1400	1500			
	Normal pressure = kPa — Flow rate (l/s/m^2) k(m/s)	(10 cm head)					
	Width(s) available × length (m)						

Note: Where two values are given, they refer to the lengthwise/crosswise directions.

	Test reference	Product reference: UCO GEOTEXTILES (High strength grades)							
		HS 120/40	HS 120/60	HS 130/60	HS 150/45	HS 150/60	HS 200/45	HS 300/45
Physical characteristics	Mass per unit area (g/m²)	300	320	342	360	377	440	625
	Thickness (mm)							
	Pore size (µm)							
Strength	50 mm strip tensile test — Max. load (N/50 mm) / Extension at max. load (%)	6000/2000 / 10/20	6000/3000 / 10/20	6500/3000 / 10/20	7500/2250 / 10/20	7500/3000 / 10/20	10000/2250 / 10/20	15000/2250 / 10/20
	Grab tensile test — Maximum load (N/25 mm)							
	Plane strain tensile test — Tensile strength (kN/m) / Extension at max. load (%)	120/40	120/60	130/60	150/45	150/60	200/45	300/45
Puncture and tear	CBR plunger test — Maximum load (N)							
	Drop cone test — Hole width (mm)							
	Tear strength — Maximum load (N)							
Permeability	Normal to the plane of the geotextile — Normal pressure = kPa, Flow rate (l/s/m²), k(m/s)							
	Normal pressure = kPa, Flow rate (l/s/m²), k(m/s)							
	Width(s) available × length (m)							

Note: Where two values are given, they refer to the lengthwise/crosswise directions.

	Test reference	Product reference: UCO GEOTEXTILES (High strength grades)					
		HS 400/50	HS 400/100	HS 600/50	HS 600/100	HS 800/100	HS 1000/100
Physical characteristics	Mass per unit area (g/m^2)	850	945	1135	1240	1550	1820
	Thickness (mm)						
	Pore size (µm)						
Strength	50 mm strip tensile test — Max. load (N/50 mm) / Extension at max. load (%)	20000/2500 10/20	20000/5000 10/20	30000/2500 10/20	30000/5000 10/20	40000/5000 10/20	50000/5000 10/20
	Grab tensile test — Maximum load (N/25 mm)						
	Plane strain tensile test — Tensile strength (kN/m) / Extension at max. load (%)	400/50	400/100	600/50	600/100	800/100	1000/100
Puncture and tear	CBR plunger test — Maximum load (N)						
	Drop cone test — Hole width (mm)						
	Tear strength — Maximum load (N)						
Permeability	Normal to the plane of the geotextile — Normal pressure = kPa, Flow rate (l/s/m^2), k(m/s); Normal pressure = kPa, Flow rate (l/s/m^2), k(m/s)						

Width(s) available × length (m)

Note: Where two values are given, they refer to the lengthwise/crosswise directions.

Product reference: UCO GEOTEXTILES (Reflective cracking and linerguard grades)

	Test reference	RC 30/30	RC 60/60	LG 10/10	LG 30/30
Physical characteristics	Mass per unit area (g/m²)	130	260	340	420
	Thickness (mm)				
	Pore size (μm) O_{90}			80	300
Strength	50 mm strip tensile test — Max. load (N/50 mm)	1500/1500	3000/3000	500/500	1500/1500
	Extension at max. load (%)	20/20	20/20	6/6	20/10
	Grab tensile test — Maximum load (N/25 mm)				
	Plane strain tensile test — Tensile strength (kN/m)	30/30	60/60	10/10	30/30
	Extension at max. load (%)				
Puncture and tear	CBR plunger test — Maximum load (N)				
	Drop cone test — Hole width (mm)				
	Tear strength — Maximum load (N)				
Permeability	Normal to the plane of the geotextile — Normal pressure = kPa, Flow rate (l/s/m²), k(m/s), Normal pressure = kPa, Flow rate (l/s/m²), k(m/s) (10 cm head)	150		150	

Width(s) available × length (m)

Note: Where two values are given, they refer to the lengthwise/crosswise directions.

		Test reference	Product reference: UCO GEOTEXTILES (Non-woven grades)							
			NW 150	NW 200	NW 250	NW 300	NW 350	NW 400	NW 450
Physical characteristics	Mass per unit area (g/m^2)		120	150	180	230	280	340	430
	Thickness (mm)		1·8	2·1	2·4	2·8	3·1	3·5	4·1
	Pore size (μm)	0_{90}	105	108	90	63	55	50	46
Strength	50 mm strip tensile test	Max. load (N/50 mm) Extension at max. load (%)	340/340 50/50	420/420 50/50	530/530 50/50	640/640 50/50	780/780 50/50	950/950 50/50	1310/1310 50/50
	Grab tensile test	Maximum load (N/25 mm)							
	Plane strain tensile test	Tensile strength (kN/m) Extension at max. load (%)	8·6	12·2	16·6	20·8	25·4	31·0	40·8
Puncture and tear	CBR plunger test	Maximum load (N)							
	Drop cone test	Hole width (mm)							
	Tear strength	Maximum load (N)	210	270	380	450	550	750	1100
Permeability	Normal to the plane of the geotextile	Normal pressure = kPa Flow rate (l/s/m^2) k(m/s) Normal pressure = kPa Flow rate (l/s/m^2) k(m/s)	300 (10 cm head)	220	150	100	80	50	50

Width(s) available × length (m)

Note: Where two values are given, they refer to the lengthwise/crosswise directions.

GeoTextiles

Quality, leadership and innovation for your most demanding requirements:

- Stabilization of paved and unpaved roads
- Erosion control
- Sediment control
- Prefabricated Drainage Structures
- Custom-Engineered reinforcement fabrics
- Tiger Drain™ prefabricated drainage structures

EXXON Chemical manufactures a complete line of woven and non-woven fabrics for a variety of construction projects.

EXXON GEOTEXTILES are available internationally and are backed by some of the industry's most experienced personnel.

Call or write for further information to

 Exxon Chemical International Inc.
Mechelsesteenweg 363
B-1950 Kraainem – Belgium
Tel: (32-2) 769. 33. 38
Telex: 24733